Metaverse

For Beginners and Advanced

A Complete Journey Into the Metaverse Virtual World (Web 3.0): Learn to Invest in NFT (Non-Fungible Token), Crypto Art, Land, Altcoin, Defi and Blockchain Gaming

Darell Freeman

Cryptosphere Academy

Disclaimer

All the information in this book is to be used for informational and educational purposes only. The author will not, in any way, account for any results that stem from the use of the contents herein. While conscious and creative attempts have been made to ensure that all information provided herein is as accurate and helpful as possible, the author is not legally bound to be responsible for any damage caused by the accuracy and the use/misuse of this information.

TABLET OF CONTENTS

INTRODUCTION

The world has witnessed several evolutions in different aspects of life, from medicine to transportation. Currently, the tech world is evolving and expanding into the Metaverse or virtual world, while some believe it is more of a revolution. Either way, something big is going on in the technology industry, which may cause a Big Bang and cause the Earth to be recreated all over. It may sound funny or scary. Will a new world be created and our present environment submerged in it? Is it possible that technology causes the generation of a new environment, which has a name already, Metaverse? Like Earth, Mars, and other planetary bodies that make up the world, Metaverse may join these bodies, and the interesting part is that it is Man-made.

In this book, The Metaverse, an insight into the Metaverse world is examined carefully. Yes, it is a worthy investment that companies are putting their millions into; however, it is enough that big companies are investing?

Chapter Two of this book discusses what you need to know about investment, the process, and projects you should pay attention to. Is the future promising, or its all talks? The Metaverse space is here to stay, and since the companies that matter are going into it, we may not have a say but be involved sooner or later.

Are there challenges this new tech space will face or is facing currently? Chapter Three sheds more light on the various difficulties the metaverse will face. How it intends to overcome or manage them, and the requirements metaverse relies on to function effectively.

The Metaverse technology will cause some shakings, and even job creation will be affected. Due to the challenges, requirements, and operations of the metaverse, skilled individuals in various aspects of life regarding technology will be in high demand. Therefore, new jobs will be created. These newly created jobs are explained in Chapter Four.

If you'd like to be relevant in building the metaverse, go through it critically, carve out a niche for yourself on skills you have already, or begin to pursue them, so have your place with the likes of Mark Zuckerberg. In addition, cryptocurrency projects in collaboration with the metaverse space have been launched. The craze for metaverse, being in touch with trends, and the uniqueness of cryptocurrency resulted in investments in some tokens to sponsor metaverse publicity and adoption. Cryptocurrency is intertwined with metaverse because of its uniqueness which will be of utmost importance to the metaverse.

The metaverse is an interesting innovation in the tech world with different opportunities. Be in touch with this space and explore its potential.

CHAPTER ONE: NFTs in Metaverse

Before we jump into the metaverse world and what NFTs mean, let's discuss something interesting we can relate to.

We all love the beautiful things of life and, most especially, riches! You may be quite familiar with the quote, "Money Answers All Things," and you could not agree less. What is life without money, assets, and comfort? Even humble people desire a level of satisfaction. Yes, you may not have it all, but your comfort is important.

Our forefathers once habited the world we live in today. Some had houses and lands; others had antiques, and some nothing. Those who had assets passed it on to their next of kin, from the second generation to the third, fourth and the transition continues. Over time, these assets appreciate doubling or even ten times more the value at which they received from their fathers. These assets are known as NFT, which stands for Non-Fungible Tokens.

Nevertheless, there is something unique about these assets to be classified as a Non-Fungible Token. Their originality and uniqueness were seen that they could not be replicated.

What if your imaginations produced results? For instance, you go to bed, sleep off, and have a beautiful dream where you won a lottery, and you're so excited that you laugh and awake to see yourself smiling. Sometimes, it may be a conversation you are engrossed with that you find yourself saying something, and in most cases, people around may attest to the fact that you laughed, cried, or spoke while you slept.

Amazing right! Now that's what it means to be in the metaverse. Better still, you can relate to having a nightmare and waking up to see yourself covered in sweat, fright, or even letting out a scream. That's the effect of a bad nightmare. A world so immersive your body couldn't tell if it was real or not, it gave back sensory feedback. This is the Metaverse.

Non-Fungible Tokens in Metaverse is like getting an autograph of your favorite celebrity like Usain Bolt and Christiana Ronaldo, their boots, face cap, or shirts (NFTs) in your dreams (metaverse). The only difference is that the non-fungible tokens can be translated into reality which sometimes happens in our dreams too.

We don't want to sound weird, but it will interest you to know that some people have seen things in their dreams that ended up real. What reaction do you have now? It simply means we all have lived in a Metaverse world but just didn't know. Some even received Non-Fungible Tokens that got translated in real life, and without doubt, you are one of them. Do you know how?

Those times as a little child, you had dreams of peeing in the toilet just to wake up and realize you've wet your bed! Familiar huh. Can we now see the picture? However, urine cannot be classified as a Non-Fungible Token because it can be easily replicated. Its composition is the same for every human; hence, it is classified as a Fungible Token.

Metaverse and Non-Fungible Tokens are simpler than we realized. At least, it is more relatable with the narration described above.

What then is a Non-Fungible Token?

A Non-Fungible Token is a cryptocurrency or form of a digital asset unique to an individual who is the owner; for example, a virtual fast and furious car exists on the blockchain; a history of transactions stored in a computer. There are diverse manners by which individuals spend their money on non-fungible tokens. The popular means is purchasing a piece of digital art and having an immutable record of ownership over it.

Just like investors, a few percentages of people who buy this digital art are not just patronizing the artist or appreciating the artwork. Instead, they believe that the Non-Fungible Token's value appreciates and be resold at a greater amount with time. So, while some simply want to boast of being the owner of a digital asset, others see it as a long-term investment. Either way, both individuals are satisfied.

The sale of artwork in March, which has no physical form by Beeple, an American artist by a primary art house for auctions known as Christie's, for almost $70 million is the first sale of Non-Fungible Tokens in history. A school of thought believes that non-fungible tokens do not exist in reality, and every individual can simply save a non-fungible token art piece by clicking on it with no need of spending a huge sum of money to own it.

In as much as the Non-Fungible Token world is promising, it is also a good breeding ground for scammers who can steal and sell digital artworks assets besides other risky things. However, there is proof that non-fungible tokens have gone beyond digital assets because they have been applied in various aspects in real life, and other establishments are digging out the potential in Non-Fungible Tokens.

As stated earlier, Non-Fungible Tokens are special and one of a kind. The features that make it so are:

-Rareness: When the demand for a product is high, and the supply is low, it means the product is scarce, and its value will increase. Normally, there is a limited supply for these tokens, and when that amount experiences a decrease, the rate of its circulation also decreases to preserve its value.

- Uniqueness: Non-fungible tokens are set apart by marks that portray their genuineness and authenticity. It also gives important information regarding the Token by the permanent mark. In school, most students have a tag placed permanently on their lockers or belongings for Identification and distinguishing from others. A mark shows ownership and uniqueness, just like a logo or brand name.

- Indivisibility: Non-fungible tokens retain their value in being an entity. That means it cannot be divided into smaller forms or parts. To purchase a non-fungible token, you must buy it as a whole or forget it, just as you cannot buy an egg without the yolk even if you do not eat yolks. The only option is to buy the egg as a whole and dispose of the parts you do not want.

Metaverse

Imagine. Okay, let's pause and take a short exercise.

Think of your favorite person, celebrity, mentor, personality, whoever it may be you are dying to meet.

Now that you have decided, what would you like to do with this person? Go on a date, sing, play with, whatever it may be.

Then, take a deep breath and see how your meeting and interaction will turn out through the eyes of your mind.

Amazing huh? The blushing made your cheeks red and your smile wide.

You just practicalized being in a Metaverse. It's that simple.

What is a Metaverse?

A Metaverse is a virtual environment where individuals are allowed to shop, interact, play, shop, etc., from any part of the globe using a virtual reality headset or google, augmented reality goggles, or any other device compatible with the technology.

Mike Zuckerberg of Facebook recently integrated metaverse to the social application, which caused a stir in the world and triggered the familiarity with the word "Metaverse." Recently, an individual spent about $450,000 to have his house next to Snoop Dogg's mansion in the virtual environment. For the sake of being neighbors in so not real life, he took out $450,000.

This is, by far, a surprising act that leaves many people wondering about the effect of the metaverse. In the metaverse, people exist as Avatars which are look-alikes of an individual in three-dimensional form. With these customized avatars, individuals switch activities easily, like sitting in a classroom one minute and the next, you are at the spa, relaxing your muscles. You can even interact with your colleagues from work or coursemates miles apart "in person" as you shop for a digital shoe.

Blockchain

The two words most likely to get conversant within 2022 are Non-Fungible Tokens, Metaverse, Cryptocurrency, and Blockchain. I'm certain you've heard these words over ten times already. It's time to be on your toes.

Blockchain is simply a virtual general ledger for assets to be traded, owned, and tracked openly without modification by persons with dubious motives. It is a decentralized system.

Decentralization

To be decentralized means transferring the power and authority from central bodies such as governments, establishments, and others to users since they have become key in today's technology. Decentralization means equal supremacy where no individual or group of persons have the ultimate power or authority over the system.

Cryptocurrency is also a decentralized virtual currency that uses the Blockchain network to track and boost its transactions.

Such change in power and influence can change many things, from the organization of industries and marketplaces to events and functions. Twitter is one

22

of these companies that intend to carry out a project to design a decentralized platform for its social users.

Decentralized autonomous organization (DAO)

Decentralized autonomous organizations are a community of internet users who own and control the Blockchain network. It uses codes smart contracts to establish the rules and regulations of the group and execute decisions automatically. This is like a team of board members where each person is a shareholder with equal rights to make decisions. However, in the Decentralized Autonomous Organization, the power of decision-making is not based on the percentage of shares an individual has. Whether you have the largest or least amount of Token in an internet community, everyone has the right to make decisions equally.

Non-Fungible Tokens and Our Virtual Lives

Suppose you see yourself in a game of virtual reality table tennis with your friends after a class in virtual school or getting hooked up on a Non-Fungible Token show about animals, regardless of the situation. In that case, it is obvious that technology is getting more real in 2022 with a touch of weird. You already made the best decision by purchasing this book and going through its pages. Technological companies such as Apple, Sony, and Meta are pushing every button and making waves to adopt virtual reality for work and fun.

What is Extended Reality?

Like the word extension, the extended reality means the process of increasing or enlarging a thing. Extended reality can therefore be defined as enlarging reality with the application of technology. It is how our present world is constructed with the aid of technology to create a unique environment for communication, creating an experience and exploration. There are three different types of extended reality; AR, MR, and VR.

Augmented Reality (AR)

This is another reality technology that is different but connected to virtual reality.

In simple terms, augmented reality is the process of placing and anchoring digital tools or objects in reality. You don't think they will appear in that physical environment because they're not. Rather, your augmented reality glasses and phone screen make it look like those objects are present in the space.

Let's take a flashback to the year 2016, when Pokemon Go was everywhere in the world as a magic spell. The Pikachu was placed on a sidewalk, usually ahead or just in front of players. That is a major use of augmented reality. Just in case you cannot relate to Pokemon Go, Pikachu's sidewalk, Snapchat should probably do the trick. The silly, funny, and beautiful filters you use to create an effect on Snapchat is Augmented Reality.

Augmented reality is yet to be maximized fully although, medical personnel can apply it in seeing the veins so that injections are appropriately given.

Mixed reality (MR)

If you understand the concept of virtual and augmented reality, then mixed reality is like combining both worlds, which you should also grasp. Mixed reality is abbreviated as MR.

A mixed-reality device permits users to immerse themselves fully in a virtual environment but view real-life surroundings in their midst with virtual tools if they desire, just like Apple's developing headset. When a mixed reality is genuine, it will enhance virtual communication I'm real environment.

Virtual Reality (VR)

Virtual reality is the application of the technology of a computer to develop an environment that is simulated. With virtual reality, users are placed in an experience rather than watching the screen in front of them; they are also immersed and can interact with the three-dimensional world. Various senses are stimulated like smell, vision, touch, etc. The gatekeeper into this virtual world is the computer.

Non-Fungible Tokens and Our Virtual Lives

Non-Fungible Tokens, as understood, are beyond artworks that are valued at exceeding amounts. The concept of Non-Fungible Tokens is ownership and value. How then does it affect our virtual lives? In easier terms, Avatar. As stated, avatars ate a digital representation of our personalities.

If Non-Fungible Tokens prove ownership, then it means every Avatar who is a dancer, fashion designer, interior decorator, or whatever it is that goes ahead to create a special design, style, music, dance, or routine, is tagged automatically with a seal of ownership. Hence, Non-Fungible Token in the Metaverse protects the rights, properties both intellectual and not, of a user, thereby regulating scamming activities.

Also, the feature of Indivisibility in Non-Fungible Tokens makes it harder, in fact, impossible for protected digital products to be replicated. We can now say that Non-Fungible Tokens is a safety technique and value upgrade for Virtual Lives.

Non-Fungible Token Technology- The Key to Metaverse

A metaverse has been defined as a virtual reality where people can work, play, and engage in activities they would love to. That being said, non-fungible tokens being unique assets, have been said to be keys to the metaverse. Non-fungible tokens are used as reward systems by the blockchain technology for games found in the metaverse. They are also identified as value systems. When an object is identified as a non-fungible token, you must know that it is of immense value. Non-fungible tokens are a paramount part of the Metaverse foundation so, being a door to the Metaverse world is only normal. Keys are used to gaining entry into a closed door. If non-fungible tokens are keys, they will be used to enter the metaverse.

Why Non-Fungible Tokens Are the Keys to Accessing Metaverse?

A three-dimensional environment that gives individuals and companies both private and public overflowing opportunities to transfer assets and products from reality is known as the metaverse. This world plays the role of offering a just and open ecosystem backed up by Blockchain technology.

Role of Non-Fungible Tokens in Play-to-Earn Gaming Sector

The play-to-earn sector is gaming technology that allows players to earn money as they play games in the metaverse. The players will be actively involved and impacted by non-fungible tokens via blockchain games. The non-fungible Token is a medium of making the seeming metaverse mountain a low land that everyone can walk through without hurdles, increasing the network of users or metaverse community. This creates memorable moments in individuals foster creativity and social interactions.

It is easy to get into the metaverse. All you have to do is purchase non-fungible tokens for in-game through the Binance Smart Chain Non-Fungible Tokens collectibles and possess access to the metaverse gaming.

The multiple application of non-fungible tokens keeps increasing rapidly; digital services and valuables have been upgraded recently in the metaverse. Digital environment the world has heard metaverse in a widespread manner due to the integration of Meta into Facebook. Mike Zuckerberg made a bold move in this launching program. Thus, non-fungible tokens incorporation with metaverse can function as the base for the evolving social medium. Non-fungible tokens and Metaverse relationships have been strongly intertwined, especially in Blockchain technology games besides various games that have become interoperable. The role of non-fungible tokens in the gaming sector has been widely accepted by cryptocurrency gamers, enthusiasts, and users to a large capacity. In 14 days, it became worth over $16 million, with every IGO of non-fungible tokens otherwise known as Initial Game Offerings purchased completely by users.

Non-Fungible Tokens Role for Trades

Digital trading has become the other of the day and the current trend. Day to day, cryptocurrency is being mined, the development of blockchain networks and technology in which wallets have also been designed in every direction. Non-fungible Token has spread its wings worldwide to the interest of several investors, which became the current trend.

Non-fungible tokens are records of different types of technology, innovation, proofs, documents of digital impacts, rare artworks, musical recordings, etc. So, each moment an investor buys the desired value, a non-fungible token is given to them as a mandatory means of proving legal rights and ownership of the claimed asset to the investor and is otherwise called a certificate, even in the real world.

When land is purchased, for instance, a certificate of occupancy to show authorization by the Government and documents showing ownership are given to the individual and kept in the owner's possession or at the bank for some. In this instance, however, the non-fungible tokens are kept in the blockchain; this is where the tokens stay, and they are immutable ledgers. These immutable ledgers regulate the non-fungible tokens accrued to the individual just like an account manager is in charge of a customer's account. The ledger tracks and regulates the tokens in the blockchain and authorizes them.

Like land, cars, and others are sold to another individual, if an individual who possesses assets in non-fungible tokens decides to revoke ownership of the tokenized asset by selling it to another, it is simple and permissible. The non-fungible Token is simply sold to the individual, and a change in ownership is affected in the documents.

The marketplace of non-fungible tokens, which is gradually emerging, is to be watched out for because it is a well of opportunities, a goldmine waiting to erupt. Trends come and go; however, certain trends create a long-lasting impact that cannot be denied. Non-fungible tokens are among the trends that will unarguably cause worldwide impacts.

How Are NFTs Redefining the Creator Economy?

Before the emergence of the Creator economy, the Attention Economy was. It was a model whereby the commodity with the highest value caught the attention and fancy of the audience because it was thought to be a rare and valuable commodity or resource. This attention economy is not new; it's been in existence for a while now. Herbert Simon, a Nobel laureate, coined the term Attention Economic. Its concept got discovered in 1997 by Michael Goldhaber, who started a shift in the economy from material model to attention-based model.

In the Attention-based model, large companies like Facebook, Google, Apple, etc., are responsible for creating what audience will consume. This is not strange in any wise. When scrolling past newsfeeds on Facebook, certain ads pop up at the side, while some even cover your phone screen in annoyance. This is not a mistake. It is a minute part of the ad plan.

When you pay attention to these ads, they automatically become revenue for the huge company involved. Social activities have become a part of our daily and normal lives, which may have negative impacts such as consuming information passively, and 'mindless scrolling,' a term used to describe the consistent need to use our devices.

This notable paradigm shift caused the attention-based economy to create a path for the creator-based economy to rule. In the Creator-based economy, the creators are normal day-to-day individuals like many of us who have taken authority over diverse online mediums to interact with people their audience.

Even though the concept is not new, the creator-based economy has shown to a large extent hoe contents are created democratically and distributed, which results in the decentralization of large, popular platforms. This has assisted creators in building alongside their passions, especially for their audience or fan base, who have rights and autonomy to decide what they want to consume.

As a result of its special setup, the creator-based economy has given individuals the right to develop digital services and products that use blockchain technology and, in time, can change or modify what this financial system will represent for the creators of the technology. Creators have also been able to gain millions from a piece of their work only. Thus, expanding the financial possibilities and freedom, every creator can attain revenue easier than in the attention-based economy where the people were used to generate revenue for a group of individuals.

The craze of non-fungible tokens has got creators involved, searching for a means to utilize Non-Fungible Tokens to share rare, paid moments and interact with their audience or fans, as reported by Sam Blake. Non-fungible tokens fix a seal on the owner; this system makes an object valuable just because an individual owns it. Relying on the fact that every human places special value on their properties, monetary or not, it is innate. A perception that is owning a thing makes it special and of greater value in itself and by itself. Thus, Non-Fungible tokens allow fans to connect their experiences with value.

CHAPTER TWO: Metaverse investment

Metaverse Boom! How can small investors make money?

The metaverse universe is going with full speed, without brakes, and it beats one's imagination. Activities like games, educational content, e-commerce, and now weddings being held in the metaverse space show the rate at which people are gradually adopting this technology and its involvement in significant moments of life. The three-dimensional space is affecting and gradually taking hold in every part of human life. Professionals state that this level of advancement is just an eye-opener, and in some years' time, tangible progress will be noted in the virtual reality world.

Metaverse space has great potential and tendencies to be known as a trillion-dollar chance worldwide in the aspect of e-commerce, advertising, digital events, and even hardware manufacturing, reported by Grayscale, a crypto mogul. Does the metaverse provide equal investment opportunities to both retail and small investors? Where are the specific places you should put your money into as investments to make it big as the next evolution in technology and the internet is being shaped?

Understanding Metaverse?

The metaverse is classified as a parallel digital universe and can be used as a game center, shopping spot, exhibition place, market, and even a virtual space for parties. Metaverse users can connect to virtual reality via laptops, smartphones, or virtual reality headsets.

Let's make an analogy between the popular Facebook social platform and Metaverse space. We currently notice platforms on Web 2.0 whereby when a website is inputted; www.facebook.com, the Facebook page appears and requires your email address or phone number and password to log into the Web and be in touch with trends or current news your surroundings. Likewise, the Web 3.0 platform needs a login detail which will be your wallet address for the Blockchain network; then, you go ahead to experience the virtual reality world.

On this Web 3.0, users can have fun singing karaoke with another person in a different part of the world or hang out with friends at the club or restaurant; these experiences give you a new, ecstatic feel of the Web. Also, a three-dimensional (3D) Avatar will be your representative instead of a profile picture as shown in Ready Player One, a movie depicting the virtual space.

A more intriguing part of the whole thing is that Blockchain technology allows you to get a piece of the Metaverse space. Lands can be purchased on the space and even built upon like a store, for instance, to sell digital products. The Metaverse space will modify how many businesses function, including their end-customer relationship, stated the COO of WazirX, Menon. It will also empower limitless possibilities due to its nature which is boundary-agnostic because of its virtual presence, unlike the physical environment for end customers and businesses alike.

Pandemic and its influence on Metaverse boom

The global pandemic COVID-19 transformed user choices for digitally assisted experiences. Menon also reported that some sectors like business offices, art, advertising, sponsored content or posts business, online education, gaming, and music would benefit from the first move on metaverse.

Major large corporations intend to create a large Metaverse environment. An example of this is Meta by Mark Zuckerberg, Walmart, Microsoft, Samsung, Hyundai, taking the lead. A change in product design and marketing will be noticed in how brands publicize their products via digital means. Under Armor, Adidas, and other brands state that Guardian Link Arjun Ready CTO is exploring the market activity to advertise their products in the Metaverse space.

Individuals will also obtain higher experiences with other technologies like virtual reality and augmented reality. The virtual space will assist in boosting the experiences of customers in various sectors, from small ones like clothing and accessories in the virtual space to complicated business deals.

How to Invest in the Metaverse

Snow Crash is a novel released in 1992 by author, Neal Stephenson, being his third novel. In the novel Snow Crash, the characters described by the author interacted in a digital surrounding that allowed appearances to be changed in split seconds, and real estate is of great value as in the real world. This digital environment was called 'the Metaverse' by Stephenson.

38

Several years later, precisely twenty-eight years, companies such as Facebook, now known as Meta and DAOs; Decentralized Autonomous Organizations like Decentraland Foundation have been working round the clock with all hands-on deck for the profitability of the metaverse to become a reality. This major act has unlocked a whole new opportunity for revenue to be generated for gamers, retail investors, developers, and digital collectors.

It is of great notice to individuals not to forget that the metaverse is in its morning phase hence, undergoing development, so the Metaverse technology value is yet to be deduced. Any form of investment in the metaverse is wise to be thought highly dangerous and speculative.

A few ways technology enthusiasts can make investments in the upcoming digital evolution are through land and stocks.

Metaverse stocks

Retail investors who desire to make purchases in the metaverse have the lowest volatile opportunity to invest in publicly traded companies. Their business plans or profits are linked directly to the metaverse. The list of these companies includes the following:

-Meta Platform Incorporation (FB: NASDAQ); Mark Zuckerberg announced in October that Facebook Incorporations would go through a primary rebranding to become Meta Platforms Incorporation. Since that announcement was made, Meta released a metaverse platform on virtual reality known as Horizon Worlds. Meta's headset, identified as the Oculus Quest 2 Virtual Reality headset, also became among the trendy and hot gifts during the holidays. Moreover, we are keeping tabs with ears to the ground to know if the increase in the sale of headsets will lead to a greater number of Horizon Worlds customers and users.

-Boeing (BA: NYSE); The role of Boeing in the metaverse is the expansion and improvement of production capacity. In an interview with Reuters, Boeing's chief engineer, Grey Hyslop, stated that plane developers and designers intend to develop a major digital surrounding where computers, robot workers, and humans can communicate effectively and make endless collaborations throughout the globe.

-Roblox (RBLX: NYSE); Roblox is the online metaverse medium that permits gamers to design and share their virtual environment with other users on the Roblox platform. Roblox was launched in 2006 and since then has grown progressively to have 24 million special digital memories/experiences, 49.4 million active users daily, and 9.5 million independent developers to a percentage of 35 every year. Irrespective of the population gathered in millions, Roblox company is yet to gain huge profit.

-Microsoft (MSFT: NASDAQ); Microsoft is still digging to discover its metaverse sector in the world of expertise. Microsoft's plans and intentions are to release Mesh, an innovation for Microsoft Team, in 2022. This additional feature to the famous video conference platform will give individuals the chance to create personal avatars and make collaborations in a three-dimension holographic environment that goes beyond the boundaries of geography or location. Holoportation is a significant character of the Microsoft Mesh design.

The holoportation is a device that permits entry to users into the former digital surrounding with a virtual headset. The user usually appears as an avatar, a digital look-alike and representation of the individual who can interact with other users and teammates like they were in reality.

41

Metaverse Land

Even though the metaverse is undeniably in its infant stage, Metaverse platforms like Decentraland and The Sandbox are already in the business of selling real estate digitally in the form of NFTs/Non-Fungible Tokens and digitalized tokens on the Blockchain network, which represents a large number of unique objects. When a person buys a square of metaverse land, normally, the blockchain network supporting the platform authenticates the purchase and change in ownership as it is being transferred. As soon as the virtual land or real estate is bought, the new owner of the metaverse estate non-fungible token can put it up for sale, rent, or even create another building on the digital property. Atari, a Japanese video game developer, bought 20 parcels of digital land recently in Decentraland and designed its cryptocurrency casino. With its local ERC-20 Atari token, bets can be placed by gamblers, and their winnings are received in cryptocurrency without taxes. Atari also publicized a plan to release its virtual hotel in 2022.

How To Buy Land in the Metaverse

Non-fungible token metaverse land is a parcel of virtual land or real estate portrayed by a non-fungible token. Based on the metaverse platform, the owner has the right to use their land for various purposes such as socialization, gaming, advertising, work, and other uses accordingly.

A Non-Fungible Token metaverse real estate or land can be bought by a land sale project or through a non-fungible token market to directly purchase the asset from the owner. To buy the land, you will require cryptocurrency and a digital wallet. Other users on different platforms can also buy land so, renting will be possible in the nearest future on the metaverse platform. When you want to buy your non-fungible token land, make sure to buy it through an ongoing project in a land sale or safely on secondary marketplaces through a trusted non-fungible token exchange platform. Ensure to understand the projects associated with the land and put the risks you are making in your financial decision through rigorous thoughts.

What is virtual NFT metaverse land?

The metaverse has grown sporadically and famous over time with cryptocurrency enthusiasts, tech fans, and investors. An interesting observation in the metaverse is the rise in virtual land demand in the three-dimensional world which has doubled, even tripled; this market has huge similarities with the real-life real estate sector. Buying and selling non-fungible land is simple and will be discussed.

A Non-Fungible Token land is a parcel or piece of land of the digital environment in a metaverse project which can be purchased. The owner of the non-fungible Token can use the land for speculative reasons only or diverse purposes. Usually, a project in the metaverse gets its map divided into smaller parts and gets it sold as a unit or various land offers. Normally, payments for digital assets are made in cryptocurrency. However, fiat currency is also accepted by certain projects. Immediately the digital asset is bought, the space offers a three-dimensional virtual experience normally for the landowner and visitors to explore the surrounding.

But some projects also accept fiat. When the digital asset is bought, a three-dimensional virtual experience is usually provided by the space for landowners and their visitors to explore the environment. Since the metaverse lands are non-fungible tokens, proving ownership and authentication is easier for the digital assets. With a third-party exchange involved, landowners can give out their lands via the metaverse project or the secondary marketplace.

Use Cases of Non-Fungible Token virtual land?

A group of investors may speculate as to what may or may not, while a few other land buyers may desire to use the land purchased for the primary purpose. However, as stated earlier, the metaverse project you pick will impact the dos and don'ts with or for your land. Normally, virtual environments host conferences, symposiums, and events and rent space for advertisement if the land generates much traffic. Several companies have made their metaverse land implemented into the services they provide. It is only normal for individuals who have bought land from a non-fungible token game if you get rewarded in-game benefits from the piece of land.

How to buy land in the metaverse

The process of buying any other Non-Fungible Token and when a user intends to buy a non-fungible token real estate is the same. To commence buying assets, you have to get a crypto wallet and buy some funds. After all, you need funds to buy goods and services. Investments are risky, so it is the user's responsibility to research before jumping into any.

A good number of metaverse mediums have designed a market for their users to sell and purchase digital land in the form of non-fungible tokens. It is easy to do this, just go through the procedures below:

1 Choosing a Metaverse Platform

-First, any user who would like to buy real estate in the metaverse is required to make the platform he wishes to buy the digital land known. Some well-known platforms are The Sandbox, including Decentraland, even though there are a few others. It would be in your perfect interest to make your findings before deciding to buy any digital land in the metaverse. The motive for purchasing the land impacts the metaverse project you chose.

2 Setting up your cryptocurrency wallet

-Next, the user has to create a digital crypto wallet; this is a form of computer software that links a blockchain network and saves cryptocurrency. It is also compatible or supported with the said blockchain network that influences the metaverse. A wallet that grants access to the crypto assets you own is better. It may either be a browser-based or mobile wallet based on your choice. In addition, you are advised to get a browser-based wallet to avoid many difficulties.

You can use the Binance or MetaMask wallet since they permit various blockchains, but you have to cross-check to make sure your wallet supports the non-fungible token land blockchain network.

You will get some Seed Phrases words when you finish setting up your wallet. Ensure to store the phrase in a secure place because it is your only means of recovering the wallet if you fail to access it in the future. It is better to write it in a book or somewhere offline like hardware.

3 Linking your wallet to the marketplace (Sandbox)

-After the wallet is created successfully, the user needs to gain entry into the market of his desired metaverse medium and the go-ahead to link his digital wallet to the platform. To find markets, go to the metaverse platform website and search.

For example, on the map in Sandbox, parcels of land to bid on that are available are seen. Some things can be done via the Sandbox market directly, while others are on external blockchain exchanges, e.g., Open Sea.

You have to link your crypto wallet before bidding on any asset is possible. Touch the sign-in section, and ensure your wallet has been fixed to the right blockchain, which is Ethereum in the case of Sandbox.

Click on MetaMask next, and a display will pop up requesting your connection. After you choose Connect, to make sure your wallet gets connected, the Sandbox platform will request for your username and email address to be added. Then, choose to continue to be done setting up the account. To use the editor on Sandbox, you can provide a password willingly. To finish your request, choose the 'sign' option on MetaMask request for signature. A successful connection will show your account details and display a picture at the top right side of the website page.

-When the user gets to this point, buying digital real estate seems so much like land being buried in the real world. Buyers are required to take note of the location, cost, and future worth of the digital asset he desires to buy.

4 Buying Tokens

-To buy a parcel of land, he has to get the coins or tokens needed to purchase the real estate and save the coins on his digital wallet. Depending on the metaverse platform chosen by the buyer, the kind of Token used to carry out the transaction differs. An example is when individuals desire to buy a parcel of land in the Decentraland platform, they will be required to buy MANA tokens since it is compatible with the platform. If the land to be bought is in The Sandbox, the person will be required to get SAND tokens or Ether for the transaction to be carried out. It is more beneficial to buy ETH since Sandbox accepts Ether only.

After the cryptocurrency has been purchased, it must be transferred into your Cryptocurrency wallet. To do this, the public address on your crypto wallet should be copied and used as the address for withdrawing or transferring the Token.

5 Choosing a piece of LAND

It is easy to go through the land for bidding or purchasing that are available on The Sandbox. If there are no available Lands, you will find them on OpenSea although, you can still make bidding via the Sandbox map. The map is the perfect way to ensure that you bought a legal Non-Fungible Token parcel of land as the OpenSea links are found in the UI. When you get a piece of land you like, you can choose to Bid so an offer will be placed on the land or purchase it for the price at which it is set by choosing the Ether amount.

Bidding

When an offer is made by bidding, a pop-up will appear on your screen, which will let you make an offer. Put in the bidding sum and select 'place bid' to confirm the transaction in your crypto wallet. If the digital land seller refuses your price or the land sale comes to an end, the cryptocurrency debited from your wallet will be refunded. However, if the fixed price is chosen, the Web will go to OpenSea to finish the transaction process. It is important to have your crypto wallet to be linked with the market before land can be purchased. OpenSea can be used to make offers if you do not want to use Sandbox directly.

-On the other hand, if the user has linked his digital wallet to the platform's metaverse market and funded the wallet, all that is required for him is to place bids on the chosen land or purchase the land at a go. The price of the digital land will then be deducted from his wallet, and the Non-Fungible Token, which represents the digital land, will get transferred into the buyer's wallet.

-To buy other metaverse non-fungible tokens like clothes and accessories, shoes, etc., for personalized avatars, apply the same procedure.

Going through the process above is quite similar to purchasing an item in reality. You get dressed and go the market you know your product is being sold, or for landed properties, you contact a real estate agent. You can even do a survey around the location desired in search of lands, when you get a suitable pick or item, you contact the seller, who tells you the cost and pay for the transaction with a currency peculiar to that environment or the seller's preference.

How Land is Sold on Metaverse

When selling your Non-Fungible Token Land, you have two options normally:

-Selling the land through the metaverse project market or

-Other secondary markets.

With Sandbox, third-party markets are the only option for sales presently. Landowners will be able to make land sales directly on Sandbox for a commission of 5% for transaction charges in SAND in the nearest future. To give out your land for sale on OpenSea, you just have to access your account profile and select the Sell option on your non-fungible Token. With this process, it will be possible for an individual to set a price or time for an auction.

How Land is Rented in the Metaverse

Certain metaverse projects, e.g., Sandbox, will provide an opportunity for digital landowners to give out their lands for rent to third parties. There is no certified system for renting land fixed already. When you make a decision to give out the land to another for rent, you have to reach an agreement privately, so the procedures are made less risky. Ensure to never make the mistake of transferring the ownership of your non-fungible Token to the renter. To be safe, be patient wait for a secure system of rentals to be developed.

Points to Watch Before Buying a Non-Fungible Token virtual land

When investing in non-fungible token land, ensure to obey the best procedures available just as any other investment is given much thought. Be certain to use the official metaverse project link to purchase your non-fungible token land or select an authentic third-party market. Before you push forward to confirming your purchase, search the metaverse platform critically by carrying out research including the platform's fundamentals. Remember that other options are available besides buying land, like renting in the future when the need arises.

Metaverse Cryptocurrency

We have read the words' non-fungible' severally in this book, and we understand what they mean. Fungible tokens, however, are responsible for powering every metaverse project found on the Blockchain network. This means that the opposite of non-fungible is fungible and that fungible tokens can be divided and exchanged mutually.

Fungible tokens are useful in purchasing digital assets and products such as real estate or clothing and accessories for avatars. These fungible tokens are also used for trading other fiat currencies or cryptocurrencies. Some specific metaverse crypto permits their owners to cast votes on making decisions in the metaverse. For instance, places where revenue should be invested or the new designs to be launched.

Digital assets increase in value, and when they do, it is normal for their supporting tokens to rise in value. In addition, certain metaverse platforms such as Decentraland ensure to burn every MANA token used to buy digital assets and remove them entirely from circulation, thereby raising the worth of the tokens left in supply. The list below shows metaverse tokens in decreasing order by market cap size.

Metaverse is a digital virtual universe that users use to access through a desktop computer or virtual reality goggles or headset and allows users to play, work, or interact in a fictional realm. Furthermore, the definition of the metaverse is based on the individual you question because we have individual differences, and they shape our perspectives.

The dictionary defines the metaverse as a three-dimensional virtual environment, particularly in a character-defining game online.

Mark Zuckerberg, the founder of Facebook, defines the metaverse as making the amount of time we spend on screens better, memorable, and more impactful. As a commitment to the Metaverse project, Facebook transformed its founding company to Meta.

On the other hand, meta virtual reality's boss defined it as a group of three-dimensional virtual spaces where immersive moments can be shared when there is no means of being together. For metaverse to be wholly immersive environments with their ecosystem, they require a currency. This is where cryptocurrency comes to play.

Metaverse cryptos can be spent only in the metaverse universe they are supported in. You purchase the metaverse cryptos in each virtual environment in-game reserve or store. The metaverse coins or tokens are used in buying in-game objects like digital lands and several products that players can trade in the metaverse. Compared to other cryptocurrencies on the marketplace, metaverse cryptocurrency values also rise and fall, which is unpredictable.

Do not forget that purchasing crypto assets and DeFi; decentralized finance tokens are dangerous to invest in. also, investing in cryptocurrency, whether metaverse or not, is not a yardstick to earn money, so be sure to be aware of the dangers involved and put your money into what you can lose only. Cryptocurrency and DeFi tokens are likewise very volatile; hence, your resources, cash, or capital investment can go down the drain and hit the sky as well in the blink of an eye. In addition, some metaverse products and crypto services are very difficult to comprehend. Therefore, ensure to make investments in things you understand only.

Major Metaverse Cryptocurrencies

The following is a list of metaverse coins that are highly risk-taking investments and speculative.

SAND (The Sandbox)

The cryptocurrency of Sandbox is known as SAND, a type of metaverse space where individuals can purchase and exchange digital assets freely. In a virtual environment run by NFT, users can create non-fungible tokens for building, owning, and generating revenue from their virtual gaming moment. Sandbox is owned in collaboration with a blockchain technology gaming developer, Animoca Brands, located in Hong Kong.

The Sandbox token SAND has a market capitalization of more than $2billion. An influential technology firm globally, Softbank made an investment worth $93million in Sandbox company when the year began. Just as Roblox, The Sandbox metaverse focuses on a user-generated platform.

Also, SAND tokens are rewarded to people who take part in the alpha user test in The Sandbox. The token SAND can be bought on digital exchange platforms also. SAND is used as a staking token, utility, and governance. People who have SAND tokens use it to buy digital products and services, cast votes on intending plans inside The Sandbox, and more to make stakings for more rewards from SAND.

MANA (Decentraland)

This is a blockchain technology-based virtual reality medium for users to buy parcels of virtual real estate or land properties known as LAND.

Later, after the land has been purchased, you are allowed to build upon, travel through, and even make money out of the pieces of land you bought based on the organization behind the technology; Decentraland Foundation. LAND is purchased with MANA, the crypto token in Decentraland in-game.

Besides LAND, the cryptocurrency is used to make payments for any in-world products and services. The market capitalization of Decentraland is around $6 billion during the writing phase. The Decentraland Metaverse is influenced by its native currency, MANA, and used it as a means of exchange on the Decentraland market.

ATLAS (Star Atlas)

Lovers of sci-fi, the science and fiction nerds should be excited about this Token, but it is one of them. Start Atlas currency, ATLAS is the currency for Sci-fi. Star Atlas is a future game for exploring space where the players take sides with factions to develop civilizations and economies in the intergalactic. Players can trade and make their non-fungible tokens within the Start Atlas environment to make money. To carry out transactions, you need Atlas; Star Atlas built-in crypto. The game is also described as more challenging.

Libra

Facebook, now known as Meta, in 2019 released a project named Libra, the platform's intended digital currency. The social network platform stated that it would like to design a single, private currency system to ensure users easily make payments across the border. Like other cryptos, individuals would be enabled to sell our base Libra tokens on exchange platforms for native currency. According to Meta, it's substitute currency; Libra will create a path for online transactions to be done swiftly. Due to opposition from various regulatory bodies, the project has been on hold for years worldwide. However, Meta is done giving up; they are pushing the project by all means. The Meta company publicized that metaverse will integrate with Non-Fungible Token and pave the way for a supporting cryptosystem.

AXS (Axie Infinity)

Compared to MANA in Decentraland, whose usage is for buying digital products and services, the AXS token of Axie Infinity is a token for governance. Individuals who have AXS tokens will have the right to make decisions on the intended plans, which will influence the ecosystem of Axie Infinity, particularly the way funds in reserve is being distributed and used. A futuristic plan is to adjust the AXS token that will be used to buy services and products on the Metaverse, Axie Infinity.

ENJ (Enjin)

Enjin is a Blockchain technology game company. It provides its users a huge amount of intertwined Play-to-Earn game memories or experiences, unlike Axie Infinity or The Sandbox, which gives its users one Metaverse service only. This Metaverse project and Token is special because the letters ENJ are fixed into every non-fungible Token minted within the Enjin ecosystem, offering real-life value for digital products and assets.

The Future of the Metaverse?

Certain companies invested in the Metaverse space greatly have spent and are still investing millions of dollars to make their customers believe that a new day is here and the metaverse is the day upon us. However, will this new day cause large adoption by people and unrestricted digital interaction or a niche service saved for future technological enthusiasts and game lovers?

Whatever happens, will only be known and revealed in time. Presently, investors in retail who are fascinated about the Metaverse space can go through those platforms and consider the future of metaverse worth by themselves.

Metaverse can be invested either directly or through indirect means. The three means of investing in metaverse directly include; buying Metaverse tokens such as MANA, ENJ, or SAND. Through the purchase of in-game NFTs (non-fungible tokens) and by purchasing virtual real estate in the metaverse space. An individual can purchase metaverse-related stocks such as Facebook

(Meta) and Apple by indirect means. Investment can also be via the Metaverse Index.

How to invest in the "metaverse"?

When Mark Zuckerberg of Facebook announced that he would change Facebook's parent company's name to Meta, other huge companies such as Disney Nike, amongst others, also began to prepare seemingly for this project that will be the next most talked about thing.

Ever since Zuckerberg made public that he put Facebook's future in the metaverse, wall street adopted it as their new trending word, "Metaverse." What is meant by the word? Meta and Universe are contractions of the word metaverse. According to Zuckerberg, he said in quote "Metaverse is the holy grail of social communication." The plan is an additional creation of virtual experiences that are immersive while the connection is maintained in real life. There are various types of virtual experiences achieved in the three-dimensional world, amongst which are:

-Purchasing clothes that are real in virtual stores and having them delivered to you in real life,

-Going to a virtual concert held by an artist in reality, etc.

These experiences show that three-dimensional virtual space and real-life are linked intrinsically. To access the different environments, all you should do is make use of a virtual reality headset or goggles and various interfaces on technology.

Also, the metaverse is an innovation in gaming systems with Metaverse projects like Decentraland where virtual users gain MANA, the native Decentraland currency, purchase land or gather collectibles, cast votes on projects in economy and administration, or even create non-fungible tokens and enhance them with interoperable features for the amount worth their time in the game. For a large population of individuals, the metaverse is perceived as an upgraded virtual reality version. However, many professionals believe metaverse could be the internet's future.

The metaverse is a virtual environment that individuals are allowed to explore using customized avatars. The environment is more than games and permits first-timers to interact and feel so many activities in real life virtually. Professionals are also in the school of thought that the metaverse now depicts a special $1 trillion resources chance.

A step-by-step guide

Investing in the metaverse can be done in three ways directly:

- Purchasing Metaverse tokens

- Buying in-game NFTs (non-fungible tokens)

- Purchasing virtual land in the metaverse, which is given out in the form of non-fungible tokens

With two simple steps, you can make investments in the Metaverse:

1: - Creating a cryptocurrency wallet

This step is basic to everyone. You need money in your wallet for every transaction and sale, just as it is in real life with physical commodities. You are to register, create a cryptocurrency wallet and fill it with cryptocurrency for making investments.

For buying non-fungible tokens, Metamask is the most famous cryptocurrency wallet. Others like Binance and Coinbase are good choices also. Most non-fungible tokens are Ethereum-based. Hence, buying the ETH (Ether) crypto with your native fiat currency is better.

Ensure your identity is verified to authenticate your wallet via completing the KYC section that is Know Your Customer.

For buying Metaverse tokens;

2. Opening an account on the metaverse platform of your preference

Buying Metaverse Tokens; A basic way of obtaining the metaverse tokens is going to various cryptocurrency exchanges such as CoinDCX and using the fully funded wallet to buy tokens directly. Amongst the Metaverse tokens which are highly sought for are MANA; this is Decentraland metaverse native currency

SAND; The Sandbox metaverse native currency and

AXS; the Axie Infinity Metaverse native currency

- Buying in-game Non-fungible tokens or owning a virtual land; you have to register an account with the intended game you desire to buy then link your cryptocurrency wallet to their account. For instance, log in to Decentraland for buying virtual real estate, Sandbox for buying and selling artistic works, Axie Infinity for parcels of land and characters, etc.

If you like, you can register an account with OpenSea to access all Non-Fungible Tokens in a mutual market for ease.

3. Choosing the Non-fungible tokens you desire to buy and make payments

If you research any of the metaverse platforms mentioned above, you will see that non-fungible tokens usually do not have an already decided price. Users have to bid for these assets and then win every other person in the bidding process to get rights over the Non-Fungible Token. The non-fungible tokens are processed swiftly via the filled crypto wallet.

Do not forget that buying from the main market like Sandboxx, Axie Infinity, etc., has its ups and downs, including other markets like OpenSea. When a non-fungible token is found on the main market, it is most likely to increase the resale value. Yet, estimating its value in a primary market is difficult. This is due to the ability to compare the costs of a Non-Fungible Token against others on the secondary market that shows all Non-Fungible Tokens on a single platform.

Indirect Investment in Metaverse

The indirect methods of investing in the metaverse are;

- Buying metaverse related stocks; Organizational stocks are involved in an active form in developing the metaverse so, they are metaverse-associated stocks. These organizations can be running networking technology, virtual reality google design, production, three-dimensional rendering applications, etc.

Famous stock choices are NVIDIA, Apple, Roblox, etc. The stocks can be bought via brokerages or the metaverse exchange-traded funds.

Investing in Metaverse Index; Just like the indices in a nation's stock market that show top-performers trends, the VMI (Metaverse Index) reveals trends in business, gaming, and entertainment going into the virtual world. The virtual world trades currently at $225.86, and the Metaverse Index lowers the dangers associated with buying Metaverse tokens since volatility is lowered greatly. The metaverse index comprises the trends in all top-performing metaverse tokens that exist.

The metaverse is experiencing an upgrade continuously, and several developments are yet to be started. Although, a great transition from the physical to the virtual world has been witnessed. There are untapped potentials and great opportunities in the metaverse, but you should research carefully before making major investment moves.

WEB 3.0

Connecting people is what the internet has always been about. In recent three decades ago, technology has changed, and our interactions with the website have transformed with this change. Three significant internet eras were in existence:

1st Era; Web 1.0 - We were connected to the internet by Netscape

2nd Era; Web 2.0 - We got connected to various online networks and communities by Facebook, including other social platforms

3rd Era; Web 3.0 - We will connect to a virtual network and space by ourselves, the user community through the metaverse, rather than the internet giants currently running the Web.

KEY CHARACTERISTICS OF WEB 1.0, 2.0 & 3.0

The second-era Web 2.0 mobile internet transformed our mode of usage and interaction on the Website. The third-era Web 3.0 has the power to generate a more impactful evolution. The opportunities analyzed in the Metaverse marketplace are estimated that virtual gaming systems' created revenue would increase from $180 billion in 2020 to a sum of $400 billion thereabout in 2025.

A significant feature in Web 3.0 is the continuous transformation of game monetization but developers. On the other hand, players are greatly switching lanes from paid to free-to-play games that have been monetized by the sale of land, players' items, and others, so their level of influence or status in the virtual world will be increased.

An important feature of evolution noticed between Web 2.0 and 3.0 is the move from closed to open metaverse on cryptography. This means from a website owned and regulated by Web giants to a Web owned and regulated democratically by users.

In the Web 2.0 universe, the player's efforts and investment in the game cannot be monetized. The game innovators in this era disallowed players to exchange items with others and ensured to close the world, so players don't get to transfer the wealth they've gathered in virtual space to a real-life economy.

However, the introduction of Web 3.0 opens cryptocurrency Metaverse technology and network to tackle this challenge by the elimination of major controllers rubbed on the virtual space by platforms of Web 2.0. Web 3.0 permits users to possess digital assets such as Non-Fungible Tokens, make trades with various in-game using the non-Fungible tokens and transporting the tokens to other digital moments, and develop a new internet ecosystem that is easily monetized in real life. This monetized creative economy is known as "Play-to-Earn."

Web 2.0 companies in existence like Facebook are mandated to adapt to and adopt the business model by changing the closed Web to open and giving out some of their unique advantages for competition. This new investment may be valuable since metaverse depicts an opportunity for revenue over \$1 trillion in creator monetization, social commerce, hardware, digital events, and advertising.

Web 3.0 Metaverse Economy

The initial Web 3.0 Metaverse world was entirely designed on the Blockchain network with numerous players who contribute to game development while permitting the elements to be traded on the Blockchain network freely.

Popular examples of commercial events currently include:

- Concert halls and event centers in the Metaverse space where musicians and DJs set up concerts and events.

- Art galleries that permit owners to showcase and sell their non-fungible tokens digital artwork at auctions like Sotheby's

- Companies that have set up a headquarters digitally in the metaverse for their workers to have meetings and collaborations.

- Advertising or digital billboards in the Metaverse space.

The web 3.0 Metaverse environment is among an intertwined economy on cryptography in the cloud. Decentralized procedures communicate and offer technical tools for the virtualized economy to be productive in the metaverse. The rate of capital investment into the sector has seen a significant increase.

Facebook, among other companies, intends to invest tens of billion dollars into the Metaverse World. Yet, this is just the tip of it for several professionals seeing that the Metaverse technology may need ten to fifteen years more to develop fully.

The Metaverse value system has been highlighted by Jon Radoff in 5 groups:

Human interface

The term human interface is defined as the software and hardware tools that allow the connection between the virtual environment and real-life environment and create connections amidst Metaverse users. Let's take a 3D Google, for example, smartphone apps, neurological implants, haptic devices, smart audio devices, etc.

This group comprises huge technology companies like Apple, Microsoft, Amazing, Alphabet, and Meta or Facebook Platform. In addition, semiconductor establishments like Nvidia, AMD, TSMC, and Qualcomm also fall in this group; other players in this niche are Equinix, Digital Realty, and Global X cyber security.

71

Creator economy

This group consists of all technology tools that permit content developers to create experiences and moments that users look forward to. Presently, certain platforms like Roblox permit content creation without any necessity for knowledge on coding. The Metaverse space needs to give expression to this sector to be developed and fully maximized to a greater extent via the usage of certain tools like non-fungible tokens, geospatial mapping, three-dimensional drawing, and so on.

The list of companies in the perfect place to fit this sector in performing a major part are Matterport, Autodesk, Unity software, Nintendo, Facebook/Meta, and Dassault Systemes.

Experience

Through the dematerialization of objects, physical environment, and distance, the Metaverse space will transform the online experiences, particularly in social networks (socializing via activities in the virtual space), online purchases (real estate, clothing, accessories, etc.), e-sports, games, live activities like music, sport, drama, etc. The experienced group has companies like Disney,

Nike, Amazon, Tencent, Song, Take-Two, Sea Ltd, Microsoft, and others.

Discovery

In this discovery, functions that permit users to get experiences on two degrees:

- Those that give support to users seeking information on certain experiences (applications, influencers, search engines)

- Marketing updates that do not match a certain search (visual ads, notifications, etc.). Companies that position themselves in this group are Facebook or Meta, Apple, and Alphabet.

Infrastructure

This is a large and wide category that consists of semiconductors, data centers, and 5G or WiFi cloud.

Metaverse Stocks to Invest In

Technological advancements have designed new companies and changed the normal human way of life unthinkable. The transistor, a part of the computer, which stands as the fundamental unit of computing, got divided into multiple microscopic parts, which resulted in a large increase of computing power, which assisted recent technologies to get grounded.

In addition, the manufacturing industry has seen notable advancements which have empowered companies to create, develop, and manufacture gadgets that people will perceive as science fiction some centuries ago. Two of these technologies that were categorized as science fiction are augmented reality and virtual reality.

Both realities serve the expansion of how users interact properly with others as well as technological services. Several augmented and virtual reality applications are simulations and video games, with both realities giving a chance for users to be immersed in a virtual technology space.

The Metaverse Industry

Another useful application of augmented reality and virtual reality headsets is what is defined as the metaverse. A Metaverse for those who have not engaged, it is an online three-dimensional virtual space that permits users to access a virtual social environment network for users to live in it—coined from the words 'meta' and 'universe' to describe a virtual communication with digital transactions and physical tools.

Irrespective of its recent fame, the metaverse space industry is said to be on a steady and powerful trajectory by multiple research establishments. A report was recently made from Reports and Data outlining the Metaverse industry readiness to raise the total yearly growth rate of 44.1% around 2021 and 2028 to reach an enormous $872 billion when the forecast session is over. The Brandessence Marker Research also reported an outline exact with the compound annual growth rate of 44.8% to attain a value worth $596 billion by 2027 end.

As a result of the seemingly optimistic speculations and its nature, both small and big players in the large technological environment have set their eyes on the Metaverse space. Many large players designing products to enable users to get the experience of metaverse or develop products that allow the great sum of computing energy needed to design the virtual enlargement of physical realities are Meta Platform Incorporation (NASDAQ: FB), Microsoft Corporation (NASDAQ: MSFT) and NVIDIA Corporation (NASDAQ: NVDA).

10 Metaverse Stocks to Invest In

Tencent Holdings Limited (OTC: TCEHY)

The Chinese Conglomerate, Tencent Holdings Limited, has been described severally by many individuals as one of the biggest worldwide game companies. Tencent Holdings possesses several gaming establishments globally; brands like Epic Games that's reached their peak of fame as a result of the Fortnite application it had. The management of Epic recently gave a hint on its desire to develop several virtual platforms for social activities that would make users immersed in the Metaverse space.

The holdings had revenue of RMB143 billion and RMB 31.75 in the total income for its fourth-quarter fiscal as estimated by analysts. A price goal of $84 was fixed by Barclays for Tencent Holdings in November, stating that Tencent Holding will have a major role in China's attempt to create its internet.

Tencent Holdings Limited is also targeting the metaverse industry, as is the case with NVIDIA Corporation (NASDAQ: NVDA), Microsoft Corporation (NASDAQ: MSFT), and Meta Platforms Incorporation (NASDAQ: FB). Product

PRX is the largest investor in Tencent Holdings Limited, with a stake of 29% in the establishment. The company's market cap hit $4.4 trillion HK on December 23, 2021. The Number of Hedge Fund Holders in Tencent is not available.

Unity Software Inc. (NYSE: U)

The company fixed at Metaverse spaces heart is known as Unity Software Incorporation, particularly when professional application and Enterprise is involved. This is possible through Unity's Software Three-dimensional platform that binds various app mediums to permit app developers to create a particular app and ensure that it can be operated on a wide range of user electronic appliances such as desktop computers and smartphones.

In its third fiscal quarter, Unity Software Incorporation gained a sum of $286 million as revenue and -0.06 dollar in non-GAAP Wall Street by surpassing both estimates. Unity Software Incorporation price goal boosted to $185 from about $160 in November 2021. During the third quarter in 2021, a total of 36 out of 867 hedge funds sector of Insider Monkey's speculation has the company's share.

On December 23, 2021, Unity Software Incorporation had a market cap of $42.4 billion. The largest investor I'm Unity Software Incorporation is a Partner group of Glenn Hutchins, Jim Davidson, and Dave Roux Silver Lake Partners. The Number of Hedge Fund Holders in Unity Software is 36.

Roblox Corporation (NYSE: RBLX)

This is among the software establishments that offer a Metaverse surrounding to individuals; Roblox Corporation. The headquarter of Roblox is located in the United States, California, and it creates an app known as Roblox Client to let users have the metaverse feeling and another app, a toolkit called Roblox Studio, which helps developers in developing apps for its users and clients.

When the third quarter ended, Roblox Corporation already obtained revenue of about $509 million and GAAP of -0.13 dollars thus, surpassing the analyst's estimation for both revenues. A price hit of $136 was fixed for Roblox Corporation in 2021 December, proving that irrespective of growth propelled by the pandemic COVID-19 in the game industry, the best moments of Roblox corporations are still in progress.

Among the 867 hedge funds, 50 were found to have holdings in Roblox. On 23rd December 2021, trade close, Roblox Corporation had a market cap of $58.9 billion. The largest shareholder on Roblox Corporation is Tiger's Global Manager, LLC, Feroz Dewan, and Chase Coleman, which hold 17 million shares valued at $1.3 billion.

Roblox Corporation was stated by Jefferies association in a letter that some companies were referred to as a Metaverse space because of their interoperability. Excellent virtual reality providers such as Roblox, Epic Games, and TakeTwo. Augmented reality is companies like SNAP and Niantic. These companies are large players in the space, although, as the Metaverse space keeps on growing, more companies will also rise to support.

The entire content in Roblox is almost generated by users completely. Roblox provides the engine that creates energy for the developer's studio, and these creators also partake of the funds that users use as expenses on the Roblox platform. Roblox intends to extend the platform, making it more educational and suitable for work envy. Several developers are working with Roblox to design a truly virtual space and presence. The Number of Hedge Fund Holders: 50

Autodesk, Incorporation (NASDAQ: ADSK)

Among the world's top prestigious software developers and providers is Autodesk Incorporation. The incorporation creates and sells three-dimensional simulator software alongside other services to experts in different industries. The Autodesk Incorporation platform enables the development of detailed and challenging virtual three-dimensional objects and structures, which allows the structures to grow and develop a metaverse.

At 2021 third quarter-end, a sum of 54 over 867 hedge funds was held in Autodesk Incorporation shares. The incorporation thereby gained revenue worth $1 billion and a non-GAAP of $1.33, which beat the estimated amount for both amounts by analysts. Autodesk Incorporation's target price was reduced from $370 to $330 in November 2021 by Deutsche Bank resulting from poor guidance and constraints in the supply chain.

The biggest investor in Autodesk Incorporation is Cantillon Capital Management, William Von Mueffling, who has a stake of $341 million in 1 million shares. The long-term development profile of Autodesk Incorporation is in place, as reported by Polen Capital. The Number of Hedge Fund Holders in Autodesk Incorporation is 54.

QUALCOMM Incorporated (NASDAQ: QCOM)

This American tech company patterns semiconductors to be used for multiple gadgets like the Internet of Things devices, notebooks, and smartphones. It is known as QUALCOMM Incorporated. The company drives to be located at the digital environment center when related to the Metaverse by giving power to the gadgets and devices that give users access into the three-dimensional environment. The semiconductor goods of QUALCOMM Incorporated are the Snapdragon chips and run hardware such as the Oculus Quest VR headset, which is important for accessing the Metaverse.

In its fourth quarter, QUALCOMM Incorporated gained $9 billion as revenue and $2.55 non-GAAP, which beat the analyst estimation. QUALCOMM Incorporated price target increased in November 2021 from $179 to $195, addressing that the 5G/fifth-generation cellular technology rollout sparks a positive boost for QUALCOMM.

The third-quarter survey in 2021 showed 67 stakes out of 867 hedge funds in QUALCOMM, who, on December 23, 2021, had a market cap worth $204 billion. QUALCOMM's biggest investor is Renaissance Technologies, Jim Simon, with $462 million stakes through 3.8 million shares based on the research by Insider Monkey.

In the first quarter, QUALCOMM was mentioned by Alger an investor letter which stated that: Qualcomm is a top semiconductor establishment with a powerful stance in the telecommunications ends marketplace that fixes the incorporation as the main benefactor of the 5G innovative network basic roll out.

It is also acknowledged as the company with the best innovative spec for 5G chips, as witnessed in their dealings with 75 primary OEMS, Apple included. In addition, besides handset, Qualcomm developed significant growth boosters such as automobiles, gaming, and industrials that offer the potential for the company to generate additional profits.

Qualcomm is a significant contributor to the economy positively through the relative returns and absolute portfolio during the first quarter in 2020, the price of shares reduced which affected its performance. The demand for chips by the market is very high yet, Qualcomm has not met its demand fully to maximize it as the production rate, and capacity is limited. Qualcomm had high expectations.

Though the quarterly survey increased and show high yields, the public anticipates a more impactful surprise positively. However, the limitations in production ability would be managed in this year's second quarter. The Number of Hedge Fund Holders in Qualcomm is 67.

Like most of these big companies, Meta Platforms, Incorporation (NASDAQ: FB), Microsoft Corporation (NASDAQ: MSFT), and NVIDIA Corporation (NASDAQ: NVDA), the company, QUALCOMM Incorporated (NASDAQ: QCOM) is also a main player in the metaverse industry.

NVIDIA Corporation (NASDAQ: NVDA)

This corporation is the main supplier worldwide in designing graphics processing units (GPUs) and goes by NVIDIA Corporation. The Graphic processing units are computing devices useful in visualizing and rendering virtual apps like video games, including other software applications.

NVIDIA has entered into the Metaverse software sector by providing NVIDIA Omniverse, a medium that permits companies to simulate products virtually and other surroundings. In addition, the graphic processing unit of NVIDIA gives energy to metaverse space since it is their responsibility to create virtual immersive experiences.

The third quarter of NVIDIA corporation witnessed its $7 billion revenue generation and non-GAAP of $1.17, which beat the analyst speculations hand down. The corporation raised its price target from $230 in 2021 December to $400, which outlines that omniverse and metaverse space has a high potential for the corporation.

A survey in the third quarter by Insider Monkey showed 867 hedge funds and 83 shares owned by NVIDIA Corporation. NVIDIA's market cap came to a whopping $741 billion as the trade ended in 2021, December 23rd. The biggest investor in NVIDIA corporation is from GQG partner Rajiv Jain, who has 15 million shares at a value of $3 billion.

The investor letter of Vulcan Value stated that NVIDIA corporation in the first quarter is a primary supplier of GPUs globally; it dominates the supply of Graphics Processing Units. The Graphics Processing Units intersect several significant computer trends like Cloud movement, autonomous or self-driving e, gaming, edge computing, processing, and artificial intelligence, amongst others. The Number of Hedge Fund Holders in NVIDIA corporation is 83.

Apple Inc. (NASDAQ: AAPL)

Apple Incorporation is the most valued company worldwide that rose to popularity and wealth through the top-notch iPhone. Although Apple does not enter the metaverse space directly, it is difficult to be aware of its plans due to Apple Incorporation's secretive work nature. The corporation has all it takes to make it big in this industry. Its features are human interaction, content creation experience, management interests, and developer tools.

A sum of $83 billion as revenue was brought by Apple Incorporation and GAAP of $1.24 for its third-quarter meeting, whose analysis was right but not for the revenue. The price target for Apple Incorporation is $165 in November 2021, with optimistic speculation on China's smartphone increased demand.

The biggest shareholder in Apple Incorporation is Berkshire Hathaway's Warren Buffett, who has 887 million shares valued at $12& billion. 120

shares were revealed to be held by Apple Incorporation in 867 hedge funds in the third quarter of 2021 by Insider Monkey. The market cap of Apple Incorporation also hit a sum of $2.8 trillion as the trade ended on December 23rd.

In addition, Apple Incorporation was mentioned by Clear Bridge in its first-quarter investor letter in 2021 stated that Apple Incorporation stocks doubled their initial investment into the corporation and is maintaining their value actively. The Number of Hedge Fund Holders in Apple Incorporation is 120.

Amazon.com, Incorporation (NASDAQ: AMZN)

Amazon is a well-known household name popular for electronic commerce. Amazon.com Incorporation connects business people and merchants globally to its clients and customers. The Amazon.com platform also pushes the company to be an important player in the metaspace since it allows its clients to browse for product catalogs virtually through augmented reality display of products that have been experimented with.

In the third quarter, Amazon.com incorporation gained a revenue of $110 billion and GAAP of $6.12 as the analysis estimated for both got missing.

Amazon.com incorporation had its price target increased to $4,500 as its positive expectations grew next year. The market cap of Amazon Incorporation got to $1.74 trillion when trading came to an end in December 2021. It was revealed by the third quarter of Insider Money that 243 people held Apples Incorporation shares from 867 hedge funds. The Number of hedge fund holders in Amazon.com Incorporation is 243.

Meta Platforms, Incorporation (NASDAQ: FB)

Meta platform incorporation is probably the most dedicated or hardworking company globally when metaverse space is involved. The company went to the extent of changing its parent name from Facebook Incorporation to Meta to portray confidence in the Metaverse industry. The Meta company intends to enlarge its network on the social space to design a virtual immersive reality that permits its users to interact with one another in three-dimension. The company also provides VR headsets for sale to provide users with an immersive experience.

The Meta platform incorporation gained $29 billion as revenue and a GAAP of $3.22. The company targeted price moved from $416 to $425 in December based on the positivity held for the recovery of resources spent on advertisements. In the third quarter, in 867 hedge funds, 248 purchased stakes in Meta company. The market cap of the Meta platform reached $932 billion on 23rd December. The biggest investor in Meta Platform is Ken Fisher of Fisher Asset Management, who holds shares of 7.5 million valued at $2.5 billion.

The amount of Hedge holders in the Meta Platform is 248.

Microsoft Corporation (NASDAQ: MSFT)

Among the biggest and most well-known software companies worldwide is Microsoft Corporation. There will hardly be any human who does not know Microsoft or at least has heard of it.

Due to its Windows OS; operating system, it has gathered widespread popularity, making it the King of this industry, dominating other products. Microsoft Corporation is developing a platform called Mesh which is to become a multiverse support system for the software team establishment, including other apps. The Microsoft Corporation also developed the HoloLens virtual reality headset sold presently for accessing the Metaverse.

Microsoft Corporation's first quarter-end was said to have $45 billion as revenue and a non-

GAAP of $2.27, which beats the analysis estimation. A targeted price of $400 was fixed for Microsoft Corporation in November, owning that the Azure cloud platform and other growth areas show positivity for the future.

The biggest investor in Microsoft Corporation is Fisher Asset Management's Ken Fisher, which has 25 million shares valued at $7 billion. The third-quarter survey of Insider shows 253 own the corporation shares in 867 hedge funds. The market cap of Microsoft also stayed at $2.5 trillion in December 2021. The Number of Hedge Fund Holders in Microsoft Corporation is 253.

Where to invest?

With so much going on in the technological world, Metaverse being the talk and order of the day, everyone seems to wonder how cryptocurrencies will be incorporated into this new world of technological advancement. People are known to be inquisitive. This nature is responsible for following up on trends and unfamiliar things. The word metaverse is famous and has stirred people's hearts to want to understand what it means.

Metaverse is a world, basically a virtual world or environment where Individuals are allowed to have fun and also work if they desire, such as

Microsoft and Windows. Other digital products of time back exist in the world already.

The metaverse space provides a new upgrade related to businesses to make use of and interact with their clients. Retail investors are quite limited in the space due to scope. However, they can purchase assets like Axie Infinity, Sandbox, etc. To explore the space and discover potentials that can be worked with to achieve greater things on the Metaverse. This world has various things in it which create euphoria, chills, and other awesome experiences. You should be aware that these coins are very volatile and are highly dangerous investments.

Therefore, every investor has to be on their toes, with watchful eyes regarding Metaverse investment and is needed to be current with the recent happenings and events in the space. According to Thakral of BuyUcoin, the Metaverse is a virtual market that requires basic knowledge of how it runs, commitment, and focus before the money can be invested into it. The most important rule of all that applies irrespective of the sector is that you should never put your money into a project you cannot bear to lose as an investor. If you cannot lose it, don't put it.

Several real estate metaverse investors desire to do more than purchase land and create something worth investing in. What the investor hopes to create will be based on his goals and motives therefore, much thought should be put into knowing the intents of the land to be purchased so that when the piece of land is sought for, it will fit into the image of your creation. The possibilities available in the Metaverse are numerous; however, there are intriguing choices currently being considered.

5 Intriguing Ways to Invest in the Metaverse

There are numerous forms when the metaverse is involved in making investments; however, for long-term benefits, the following ways should be put into consideration.

Large Crowds through Metaverse Event Space

An amazing way of making money in the metaverse space is by giving out your land for rentals as an event center for short-term bookings. You will be in more luck if your land has a large capacity to contain a great number of people for virtual shows, live bands, concerts, etc. Because attracting top artists, performers, and the likes who look for new means of reaching out to their fans and audience who have not been able to attend live events.

As you do this, ensure to create a space for retail. Events like these come with a lot of noise and swags; face caps, glasses, hand bands, customized shirts, etc., can also be sold. Everyone likes to show off. They went for a top event, more so if you can take a digital object as souvenirs. The thrill, yeah!

Huge Billboards and Banners like Real Ones

Advertising is an old marketing strategy; the billboard stands the test of time and is still as effective as always. Everything related to adverts traffic in real life also occurs in the Metaverse space. The lights, toads, and traffic that can be thought of as pedestrian road traffic exists in the metaverse platform. Billboards are essential platforms whereby a great population of people of different races and cultures who exist in the metaverse is reached out to.

Like the real-life space, billboards generate income for Individuals who give out the land for renting space to various companies who desire to pass a message across to passersby. Look out for virtual and real-life companies, for they are potential customers for the Billboard advertising sector. Who knows? You may get contacted to release your land for some cool cash.

Wild and captivating experiences in the metaverse

There's so much thrill and excitement to experience in the metaverse space. Did you just say Amusement parks! Don't think twice. The excitement waiting for you in the metaverse is nothing compared to what you may have felt on a roller coaster heart-pumping ride. If you think metaverse has nothing to offer because gravity and physical presence is absent, then you are so wrong.

You have to create something different, huge, and catchy so that your tickets sell quickly due to built-up anticipation. Metaverse is all about social moments and experiences, and you're hitting its purpose this way. The real-life parks have the same thing for donkey years, but in the metaverse, your imagination can be let loose always to modify, update, or remake the experiences users can feel. The limitations of real-life amusement parks shouldn't let you waver.

Boutique, malls, and other social centers are still needed

Being a virtual space does not make it any different from the real world because people are looking for activities to populate their virtual lives. The place of retail in the metaverse space is powerful; showrooms and shops are still very much needed by brands for their businesses. Both small and big brands need a place to let the world discover their potential. So, as you plan to lease out your land for such purposes, conquer your fear and be bold to reach out to prospective tenants and sell your land (we mean to sell your idea, not figuratively). A well-designed and beautiful rental environment catches the eye of every business person.

Also, many brands have witnessed tremendous success just in selling and publicizing non-fungible token versions of real-life goods to the denizens in the metaverse. Products like high-end bags, shoes, accessories, customized sneakers, including other objects, are hard to obtain, which can help in helping customers be more expressive.

5. Office space are also a great investment like the real world

The pandemic taught the workforce and companies that it would lead to havoc without effective online means of communication when another pandemic strikes again. As such, a means of unifying teams in a virtual space is sought by various companies. Hence, the metaverse office environment on large platforms instead of depending on committed metaverse is the solution to add an extra feature to the virtual space in real life. It is more fun to build a team and interact on the metaverse.

In addition, virtual designers, architects, realtors, and likes significant to metaverse development will also require space on the platform to work. These people are in existence and are getting into the virtual space to provide their services and earn money like they do in reality.

Significant Points to Note

A Metaverse is an energetic place with diverse people and needs therein; it is not a monolith. Investors in real estate are discovering that there are various aspects of realty structures needed in the space since there are various choices and different purposes. Choosing what aspect of real estate becomes a challenge for most investors.

CHAPTER THREE: Challenges of The Metaverse

Several technological news has been overwhelmed with updates about the metaverse. The noise metaverse created is second to none and obvious to every person who cared or not. In as much as the word metaverse is not new and has been in existence since 1992, the recent buzz it has created made people curious to know about this world.

According to Oxford Dictionary, the metaverse is an environment that is virtually real whereby users can interact with an environment generated by computers and users in the simulated space. What can we say is the cause for the metaverse's sudden interest, especially at such fast moves?

The sudden interest seen about metaverse, its popularity, and expansion is connected to the technology giant or the man, Mark Zuckerberg publication to modify the social platform, Facebook, into the next metaverse. This goal is said to make the generation of a single virtual world that connects creators, products, entertainment, workspace, commerce, community, and many others.

How far has it come in the realization of this vision? We may most likely be nearer to getting submerged into the virtual world than we expected. What will be your reaction when you awake on a certain day only to discover that you slept in a virtual world and your eyes got opened there also? The rate at which headsets have been on the increase and the headset project by 2024 is to go beyond 34 million units.

The virtual reality hardware developers and pioneers are Pico, VIVE, Oculus, and the likes. On the other hand, software pioneers and developers are being aware of the endless opportunities in virtual reality ending up as a worthy product in the future. The Chief Executive Officer of Unity speculated that by 2030, VR headsets would be as common as game consoles.

Presently, it is becoming undeniable that many industries are getting their technological adaptability accelerated in using virtual reality. 2021 introduced several virtual reality headsets for consumer use and business, ranging from the likes of Huawei virtual reality Glass 6 Degree of

Freedom Game Set, Pico Neo 3, HTC VIVE virtual reality, and HP Reverb G2 Omnicept Edition.

The Metaverse is indeed a promising space; however, like any other ecosystem or world, it is bound to face its peculiar difficulties. For now, we still have a pretty vague idea of what the actual Metaverse will look like; it is in between, a gray sector neither good nor bad although, it is located in a region where it can be both harmful or of advantage to its users. Yes, the Metaverse is in its morning phase, but to make it a full-fledged virtual world, we can imagine it to have to overcome the following challenges:

The Challenge Of Reputation & Identity In Metaverse

When the real world is involved, representation and self-identification are quite straightforward to answer and easily related. However, when the virtual space is involved, or even the metaverse, the curiosity that comes with what makes up a person's identification, especially how one can prove that they are the same person, rather than impersonating another or your existence being copied or mimicked.

Certain questions such as "Are you the same person with your avatar?" "How can we believe that it is you? Besides related ethical challenges, verifying identities is faced in the metaverse. At this point, a person's reputation seems to be very significant in the aspect of proof and authentication that the entity an individual is interacting with is authentic and trustworthy. The primary issue faced by the metaverse is the possibility of forgery; of footage, voice, facial features, and other aspects that may want to be copied; hence, a good number of authentication techniques will be created and implemented into the technology in days to come.

The Challenge Of Data & Security In Metaverse

Although many industries, organizations, and companies have been upgrading the IT security of their systems, the issue of security and privacy of data continuously remained an issue of worry for the users of online platforms. The manner of data application in the metaverse? If metaverse is a new plot by huge technological giants to generate extra data?

The worry is sponsored by the thought of how the virtual environment can boost a place where data abuse, as well as incorrect information, is the order of the day. Another issue is the fear of centralization is when data is regulated. The probability that the present form of security may not sustain the Metaverse.

Entering the metaverse at deeper levels of immersion requires upgraded security techniques to a greater extent of being at the equal plane as the metaverse, which is expanding and continuously demanding. This advanced level of security needs newer techniques to be developed to protect the privacy and personal data, which can certify the security of a user's identification and properties in the virtual universe. With this carefully pondered on, the metaverse space will most likely get to a stage in the future where more information on users has to be provided compared to the data needed today for self-identification and ensuring the safety system is functioning properly to keep user personal information secure.

The Challenge Of Currency & Payments Systems In Metaverse

For many decades, digital currency has been a part of our lives, particularly with the Bitcoin buzz, which is presently the most popular. Even online markets where a large population of customers in their millions get connected worldwide, such as eBay and Amazon as the popular ones.

Without doubts, the version of the virtual market peculiar to the metaverse will be designed for supporting and linking various types of digital currencies for swift transactions and stress-free exchanges. Irrespective of currency and market formation, it becomes necessary to create a unique system of verification for authenticating transactions when transactions are involved. The challenge shows making users believe that whatever transaction taking place in the metaverse can be trusted to be safe and reliable.

The Challenge Of Law & Jurisdiction In Metaverse

The metaverse is a virtual environment where many real-life activities are copied like going through goods and services and buying them. Just like other virtual realities, metaverse brings togetherness amongst people. Though it provides a chance for connections and healthy relationship building, it makes them open to various issues. Where there is no law, the people perish.

This recent metaverse technology will attract a large population of users, thus becoming a space where life-changing opportunities for interaction and trades. Users are easily susceptible to various dangers since the laws regulating boundaries are absent. It is likely a very challenging issue to see a legal jurisdiction including a jurisdiction that will make sure the virtual environment is well-secured and safe for metaverse users.

The issue of owning intellectual properties can also occur in the metaverse. In this case, when Artificial Intelligence designs a product, they will most likely not get rights to protect that intellectual property because works are said to be copyrightable only if it generated by an individual (person, human). Another issue is content creation. These creators will find it challenging to safeguard their properties since tracking copyright infringement in virtual space is hard and yet to be worked on.

The Challenge Of Ownership & Property In Metaverse

The subject of one virtual world, one community, one goal is likened to the real-life environment where interactions between the users and environment occur. There will also be a chance to buy and keep properties of different objects and valuable assets. Many people are purchasing properties and non-fungible token arts, which has

103

increased the growth curve of tokens and assets in the virtual space.

Non-Fungible Tokens popularity has doubled, and the pace at which it is making giant strides is overwhelming appealing to extra investors and virtual users to digitalized tokens and assets. In similitude to how non-fungible tokens depict real-life items that prove and give the right to ownership for videos, music, and art, it will be more challenging for a unified metaverse universe to be created and used to verify the owners of digital assets in the space, metaverse. Yes, every individual in the Metaverse is allowed to buy and have ownership over digital assets; two primary issues that arise from this are;

- How ownership privileges will be given to creators of digitalized assets

- How these digital owners will be authenticated.

A major challenge with the metaverse space currently is the creation of fragmentation. When the ecosystem gets fragmented, it generates experiences that hinder what a truly immersive virtual experience should be.

The Challenge Of Community & Network In Metaverse

Every project, especially new technological innovations and cryptocurrency platforms, has one goal: creating a community and bringing people from various tribes, regions, and continents. The goal is the same in the metaverse that a unified digital space accommodates everyone.

From time past, building communities and the various group of people that exist in our real-life environment has proven that being together and connected is important for society since we need each other to survive. When existent in the metaverse, this community creates a strong bond for users in the workplace and individuals who have a common goal to achieve.

Many people will create important relationships in the metaverse space, just like some were created on the Facebook Platform where Individuals in different continents interact with one another to the point of developing feelings and getting married. For some, it was a healthy friendship, and other opportunities resulted from a person on the platform.

The online relationship and interaction are not new but, Metaverse is unique in that it creates an environment for Individuals to be present and connected both physically and emotionally. For this to be achieved, motion capture technology alongside haptics has to evolve totally to a new height. This height and development create an easy understanding of environments and presence and visual trust that comes with sensory and touch abilities.

The Challenge Of Time & Space In Metaverse

It is common knowledge that things that happen in real-time are different from the virtual space, where accurate time perception is absent since the users are not conscious of their bodies due to immersion in virtual reality. Therefore, when total immersion is attained, it results in more session time by users in the Metaverse environment because of the perception of time which is now distorted.

It now becomes important for a means that users will be in tune with real-life to be created. Also, the element of space should be carefully thought of in the Metaverse universe. The space the Metaverse occupies is infinite, which is most likely to affect users getting into the space initially because they will not likely understand the extent of the environment as its extent is unknown.

The element of space alongside time concept in the virtual space will need users to be guided as they get submerged into this reality for the first time, which will make sure they will be comfortable and conscious of the virtual space.

CHAPTER FOUR: Investing in Metaverse; New Opportunities in Virtual Worlds

Yes, virtual space is at the morning phase. Setting up the Metaverse will have loads of challenges, as discussed above, and opportunities simultaneously. Without disbelief, it is important to ensure that the Metaverse changes into a substitutionary or, better still, a complimentary environment that does not replace our real world.

It is good that we anticipate the virtual environment, but it is necessary to be aware of safety and security in the Metaverse space. For a virtual environment that provides unlimited experiences to be created, the definition of Metaverse by Matthew Ball is to be a community of interoperable experiences.

A metaverse universe that is understood has ecosystems in fragments that are bridges for creating a virtual surrounding mainly featured by interoperation among singular components or sectors. The virtual environment recognized as

Metaverse may become tomorrow's world; however, the technologies supporting it currently provide some opportunities for the Metaverse investors.

Human life and experience have always been spurred by a passion fueled by curiosity to explore environments different from ours in every way. Experiences such as traveling to other existing planets besides Earth, traveling across the world, and unfamiliar places sparks human imagination and boost creativity and investment. As such, it is no different from journeying into the virtual space. This is Metaverse.

It will be easier to see the metaverse universe as the future's digital world, more like the next big bang the internet will experience. When it has been fully maximized and upgraded, individuals will both play and work in totally immersive worlds. The fact that key players in technology, like the metaverse idea, even go as far as investing their companies in it provides investors.

These two technologies are virtual and augmented reality (VR and AR). They are evolving to generate more digital connection experiences. Thus, the sum spent in VR/AR all over the globe in

2020 is about $12 billion, to an estimation in 2024 of $72.8 billion.

A more detailed view of how virtual reality and augmented reality technology are applied presently and how investors should partake in this shift in tech.

Beyond gaming: New Applications of Virtual Reality and Augmented Reality Technology

The most common uses of virtual reality and augmented reality technology are entertainment and games as virtual reality headsets and augmented reality glasses. However, these forms of reality are also used by concerts and events promotions that hold virtual musical performances live and museums that offer virtual guides that show a mixture of the real world and virtual space.

Besides the entertainment sector, virtual reality and augmented reality will be used in commerce actively in cases like virtual trying of clothes by customers or going through an automobile before buying it.

In the Business sector, companies that deal with large pieces of machinery and hazardous or toxic workspace use virtual reality and augmented reality for training their workers on operating the machines and safety measures.

For Health Care and Medicine, virtual and reality have been applied in emergency wards and intensive care units, particularly during the COVID 19 pandemic, to get extra professionals in the units without the dangers of being exposed to the virus.

Another interesting sector of augmented and virtual reality is used in the Office. Several companies have their initial interviews in digital environments, which lets them streamline the process of employment and increase the interview capacity to accommodate more applicants.

The Path to Greater adoption of Virtual Reality and Augmented Reality

Virtual and augmented reality adoption is beyond customers only because other industries, sectors, and companies are beginning to adapt to their realities. For virtual reality, augmented reality technologies to be successful and the metaverse to fully realize will be based on technological advancements and acceptance/adaptability of users.

On the aspect of technology, some difficulties arose, such as the cost of hardware being high, yet it is a necessity to have limitless virtual experiences, accessories, and tools that gets movements captured, committed devices like headsets and goggles, with complicated processing capability and adequate energy to give support to the virtual universe.

Also, large sums of data have to be transferred and stored through wireless means, and metaverse users have to be connected from any part of the world they are in; hence, the coverage is expanded, and the use of ultra-high-speed for the internet will be a necessity.

The zeal and willingness of consumers to enter this virtual universe will determine the rate and volume to which it is accepted and adopted worldwide. One significant question that keeps on pondering on several people's minds is whether the apps are useful as said and if all will easily access them. Will the Metaverse universe provide authentic, expressly communicated advantages and contribute excessively to lives?

Means to play the Metaverse universe

The manner for investors to get exposed to the current technological advancements and the increasing hype gotten from the metaverse leaves them wondering. Several sectors' investors can consider includes:

-Gaming Sector: Research by Morgan Stanley in 2020 estimated that COVID-19 generated lockdowns led to a sporadic increase in-game adoption over the worth of four years in the areas of the player population, in-game revenue increase, and session time while playing games. In addition to the increase experienced in the gaming sector, the choice of the social media platform for various individuals, including mediums that provided multiplayer online experiences, the aspects across social media, games, entertainment also the virtual surrounding got blur.

One thing game innovators and developers will not do is waste the chance to transform in various types of connection and maintain the content and experiences of their in-game fresh. The gaming sector is a popular trend whose adoption is outsized in 2020.

-Consumer technology or communications Sector: Many technology companies are making products for users to interact in the virtual universe effectively. Imagine virtual reality headsets and augmented reality filters that can be doubled with a camera from a smartphone, including personal computer operating systems which has more connections. The platforms for streaming and interactions can also boost the technological transition by providing content in the virtual environment than screens.

-Cybersecurity/digital infrastructure Sector: Various important innovations and trends are advantageous from the adoption of virtual reality and augmented reality. Amongst these benefits is the Cyber security system. The more immersive and digitally linked a world, the more likely cyber scams and threats will occur, and susceptible people will become. The digitalized boost will need a greater speed for wireless connection sponsored by the 5G network, which means investing in 5G digital infrastructure and mobile connection.

Whatever investment choices made by you must reflect on the goals, risk tolerance, and timelines set by you; however, for technological investors desiring to maximize the virtual environment, the Metaverse technology has exciting offers in-store.

Metaverse Jobs That Will Exist by 2030

If we journey six years back, it is 2016, the year when Pokemon Go was all over the world, making many people feel like we were on the verge of having a revolution of augmented reality. The time for its materialization did not come yet.

Coming back to the world today, we are still on the same topic on the Metaverse and if the time is now seeing that the social media platform Facebook, became Meta and the company made a mind-blowing investment into the creation of an immersive digital universe for everybody to reside. With this investment by Meta, the dream of one man will gradually become the dream of the world irrespective of our opinions and choices because the key players and technology lords want it.

The Metaverse investment and technology has caused the creation of several jobs but, before we run through it, let's revise the following terms:

-**Virtual Reality (VR):** This is a completely artificial surrounding fully immersed in a virtual space.

-**Augmented reality (AR):** In the augmented world, virtual objects or tools get overlaid on the environment in the real world; those digital environments then improve this real environment.

-**Mixed reality (MR):** When combined with the real environment, the virtual space gives you a mixed reality. In mixed reality, interaction is made with the real-life environment and virtual space.

-**Extended Reality (XR):** This is also known as the Metaverse; it is a mixture of every reality noted above.

Metaverse Research Scientist Job

Augmented reality and virtual reality researchers are amongst the employees in major technology companies and top institutions. However, the Metaverse or extended reality, a world where both digital and physical worlds intertwine to form a seamless experience for individuals, is gradually becoming an idea that is accepted all over the globe hence, the need for more intelligent researchers and brainpower.

The job description of a Metaverse Research Scientist does not go down to designing some digital objects in reality, wherein large companies will get more partners and clients. That technology already exists. The future of metaverse is bigger.

What metaverse research scientists will have to develop is much bigger.

It is related to Metaverse theory, where the whole world is digitally active and clear, just like the movie Ready Player One without the excitement and thrill. This technology built by researchers will become the basis for other applications such as Decentralized Financial systems, games, good control systems in factories, health care, and many others.

The job taken by metaverse researchers is very complicated. Individuals who would desire this responsibility have to learn to build, scaling prototypes with the use of technology as well as the integration of algorithms in computer vision for three-dimensional automated photography, scene reconstructing, sensor fusion, state speculation, mapping and localization, neural rendering, automatic imaging, and many other prototypes that will have to be upgraded as time goes on.

To become a metaverse research scientist, you need a Ph.D. in computer vision, deep learning, computed imaging, or computer graphics. Having a good understanding of C++ is also important.

Metaverse Planner Job

It is easy to have ideas and thoughts; however, executing them is expensive. When the metaverse is functioning effectively, planning and implementing any productive thought into the virtual space is of great importance to many companies. As much as implementation is necessary, choosing the right decision is also important in enlarging the digital space.

A metaverse planner will be highly sought at this point. As most Chief executive officers present a mission and plan for their businesses to grow and generate revenue in the metaverse, the metaverse planner has to push a precise portfolio in achieving goals from the deployment of proof-of-concept strategy to pilot strategy.

In simple terms, this is the identification of opportunities in the market, upgrading key metrics, developing business cases, affecting engineering roadmaps, and so on. All the fun regarding plans and strategy are handled by the Metaverse planner.

The job description may not be pleasant to the ears; however, it is critical. For instance, what will be the best advice to an automobile company? Will it concentrate on generating virtual test drives or implementing a digital twin project to foresee breakdowns? The answer is certain, unknown. Yet, it is as simple as ABC for a Metaverse planner to figure it all out.

To become a Metaverse Planner, you need many years of experience in management, an exceptional mindset in entrepreneurship, and knowledge of diverse marketing techniques; HW/SW/SaaS/PaaS.

Ecosystem Developer Job

You need to understand that the metaverse universe will not exist independently. Even the world we live in has an ecosystem that maintains it. Central processing units, graphics processing units, know your customer procedures, sensors, laws, regulations, edge computing, data-lakes, green electricity generation, and the universe is complicated, and making it more digitalized is not simple.

This is just like an automobile company transiting from fuels to electric cars. Yes, the products are available; however, the largest hindrance to the widespread adoption of electric cars is the absence of charging shops on the roadside and in the geographical location entirely as well as the consistent upgrade in battery capacity. Likewise, a product's software and hardware devices can be available in the Metaverse and yet lack other things.

An Ecosystem Developer is responsible for the coordination of governmental bodies and partners to make sure the numerous functionalities generated are on a huge scale system. The ecosystem developer will thrive for infrastructural investments by the government and a great community of animate actors.

A major thing the ecosystem developer must focus on is the key element; interoperability to make sure customers in the metaverse can make use of their virtual goods through various experiences. There will be no need to have great hair in a game if you cannot go clubbing with it. In addition, extra efforts will be given to financial bodies that require to back up their ledgers and contracts for products and services to get traded on the metaverse universe.

To become an Ecosystem Developer, you must have years of experience in Governmental/lobbying and an in-depth comprehension of the burgeoning extended reality industry.

Metaverse Safety Manager Job

Everyone desires privacy, even on the internet. We cannot say for sure that it has proven to be a secure environment for us all. Hence, whoever thinks the metaverse universe will be more effective is joking. Without a doubt, the metaverse has streams of opportunities embedded in it to be well-secured, but the safety we desire in the metaverse has to be spearheaded by people.

In the aspect of in-world identification verification, privacy, efficient sensors, secure headgears, digital tools, etc. People who can offer guidance and great oversight for these aspects during validation, mass production, design, ensuring the digital environment is secure, meeting the safety requirements.

These have to be sacrificed with good precision, maintaining revenue, keeping excellent designs, and others. The individual who can do all these is best for the position of Metaverse Safety Manager.

The job descriptions already scream difficult. The metaverse safety manager needs to precisely predict how metaverse functionalities will be applied, abused, identifying safety key parts and systems, including creating steps in tune with precision made. Just thinking of the sheer complication and the number of moveable parts in a digital space is challenging.

To become an Extended Reality Safety Manager, you need. A degree in engineering and experience in manufacturing or consumer electronics.

Metaverse Hardware Builder Job

Codes are not the only things the metaverse will be built upon. Other foundational blocks needed by the Metaverse include; headset, sensors, and cameras.

- Headsets; are used for additional real-life simulation to let you feel the sun over your head as a summer season is projected.

- Sensors; allows you to feel the sense of touch; therefore, when a person grabs your arm or kicks your legs, it has the same impact as the reality

- Cameras; enables the graphic processing unit to read your mood to know if you can be engaged in an activity or the artificial intelligence should let you be.

That sounds fun until the not-so-fun parts like localization, visual light and depth cameras for assisting tracking, inertial units of measurement, and mapping.

So, you know, the listed hardware required to generate a digital universe that mixes with the physical environment is costly and complicated; hence, a Metaverse Hardware Builder is saddled with the responsibility of assembling the hardware and adapting or upgrading it as the complexity of the Metaverse increases.

The best sensors in existence were designed for automotive companies and industrial establishments. These industries have an excessive sum of capital. An extra challenge to the metaverse builder is that he needs to ensure those tools do not need many resources to be built (cheap), and it is safe, so the metaverse universe does not become a plaything to the wealthy.

To become a Sensor Builder, this is quite funny and challenging, but you need to own a factory that can design complicated consumer electronics.

Metaverse Storyteller Job

As the growth of the various forms of extended reality increases, many people have experiences to share on their first encounter in the metaverse space and subsequent interactions. Even gamers are getting more ideas on new gaming concepts. The evolution of these sectors should not go unnoticed. Such a revolution needs a story that can be told to generations unborn and where great lessons can be picked. We have a lot to see, feel, learn, explore, react, etc. The Metaverse storyteller is just the one you need.

A metaverse storyteller is responsible for creating immersive scenarios and quests to be explored by metaverse users, training locations for military forces, difficult to pick marketing offers as narratives needed in corporations, bouts of psychological sessions, and so on. Although, the Metaverse storyteller payment is heavy not unless you see the bigger picture in how they appeal to others through narratives. It is a dream to see others motivated and escape the hardships of their day in the metaverse.

To become a Metaverse Storyteller, you need a Major in literature and a minor in Marketing. You should begin your career at a gaming company, after which you switch to Tech.

World Builder Job

The entire universe also needs to be created as soon as the architecture is built, same with the hardware and storylines. The position of a World Builder needs similar skills as video game developers, although with a unique guideline. These world builders have to think and face forward because most of their dreams will take a long while to be created as tech innovations or product answers.

Also, ethics, guidelines, and rules governing the Metaverse world have to be considered because when the realness of the digital world cannot be denied, would it be okay to murder it? Would it be okay to also commit crimes? Can a person commit suicide? These and more questions occupy the thought of various tech enthusiasts.

To become a World Builder, you need to be skilled partially as a warrior-poet and a graphic designer. Knowing Minecraft also comes in handy.

Ad-Blocking Expert Job

Have you been wondering how Facebook or, rather, Meta generates income? Then, you are not alone. Is it through the sale of subscriptions to misinformation companies? Or through donations from people? The harvest and sale of organs? They are all wrong. Meta generates income by selling ads.

The metaverse may function in the same manner. Some may say it is DNA. Do you feel like the ads on Instagram are annoying and precise? You just wait till you observe what they will most likely do with a world of data and can now track your every movement all over the world.

Let's say you are taking a stroll through a street in the digital environment, and you suddenly feel hunger pangs in reality. Unconsciously, your gaze intensifies as you approach diners, restaurants, and takeout spots during your walk. Then after a minute or two, you begin to get food ads. I'm sure this sounds exciting initially, but as it goes further, it gets intrusive.

In no time, this novel act is exhausting, and users seek ad-blockers which has to be very advanced to recognize ads that are in real-life. An Ad-Blocking Expert steps in at this point. In a similar model with AdBlock Plus, plug-ins that hinder ads from appearing should be developed. As much as Ad-Blocking Experts won't get much pay, they will earn a living by donation and data access or use it as a secondary job.

To become an Ad-Blocking Expert, you must have a basic understanding of coding, including access to the source code of the metaverse universe.

Metaverse Cyber-Security Job

The metaverse hype has made it an ideal target for attacks and fraudulent activities from the internet, such as non-fungible token theft, hacked avatars, headsets, biometric or physiological data leakages. There are just many things that could go wrong, hence, a Metaverse Cyber Security Professional.

These experts prevent real-time attacks and make sure laws, guidelines general protocols remain amended and fixed, including all dangers of the metaverse universe. The extent to which breaches may occur in the virtual space may leave us witnessing court cases in reality over virtual breaches.

To become a Metaverse Cyber-security Expert, you need a background in basic cyber-security and a degree in law with Technological inclinations.

Unpaid Intern Job

This job position is not new. I agree; however, we need to highlight the importance of this position as it will keep on being key to the future of metaverse. Who are the Unpaid Interns? They do not get coffee alone but bring out the data. They are responsible for writing codes for the trees, water, and rocks; they also create appendices for VC decks. Unpaid Interns are the important fodder on whom technology empires have been raised. They should be appreciated and paid also.

We will experience comfort, rumble sensors or beta, security, portals for punishment and pleasure, engaging narrative, optimization, customized hairstyles, ideological compatibility, parental controls, notifications, puzzles, widgets, add-ons, and many others, all within the metaverse space. However, we will not experience all this at a go.

The Metaverse space naturally needs many companies, recent technologies, innovations, discoveries, and protocols to function. From this description, it is obvious that many hands have to be on deck. Countless responsibilities, requirements, and administration have to be in place for the metaverse to work. This means people are needed, and the gradually rising metaverse world will come crumbling without the right hands.

Aside from our already existing universe, Earth, a new universe was emerging and called The

Metaverse. It is a reality that seeks to redesign every activity we engage in physically into the digital or virtual realm. We have stated over and again how Facebook's change in name is greatly connected to the ongoing metaverse project and its impact on it. With Avatars, people get represented in the virtual world.

Avatars can therefore be used for any activity in the metaverse space ranging from shopping, clubbing, working, sleeping, marketing, etc. It also fosters communication between work colleagues to interact outside work subjects and know each other better. Also, partaking in social activities such as concerts, fairs, events, symposiums, job interviews using augmented reality.

Furthermore, tech experts and business people will be required to constantly invest in themselves by acquiring skills, knowledge, social dynamics and communication, and others as the Metaverse and its technologies will also spread widely and be accepted.

Top 5 Blockchain and Crypto Projects

The rate at which the Metaverse universe is growing is rapid. The Metaverse project is increasing continuously and bringing connections to assist in bringing additional areas of digital lives in unity. Blockchain technology has a major role to play in the development of the metaverse.

Binance SmartChain is the arena for a great populace of metaverse space projects. Some of these metaverse projects are games such as Alien World, Cyber Dragon, Second Live (the other Metaverse universe), and also a Casino player-owned game known as decentral games. TopGoal, a collectible card game, also has a spot in the metaverse.

The Blockchain network Ethereum has the Sandbox and Decentraland metaverse projects that create the same metaverse space for users to create a means of digital identification, buy land, and trade non-fungible tokens on the Non-Fungible Token market. These metaverse crypto projects combine play, work, and life to the extent of allowing players to partake in the Play-2-Earn decentralized finance economy.

When users interact and play the game, they can earn income. Bloktopia also offers a similar metaverse experience spread through the 21 steps of a virtual skyscraper. Users can therefore make trades and lease out space for real estate on every floor to create a source of income.

Although the Enjin metaverse crypto projects do not provide a three-dimensional virtual reality space for exploration, it offers certain tools for generating in-game non-fungible tokens assets. Non-Fungible Tokens are also significant aspects of the Metaverse universe since they generate digital collectibles. Users of the space can design liquid non-fungible tokens that can be disintegrated into Enjin tokens at any time through Enjin.

Introduction to Blockchain and Crypto Projects

The year 2021 has been an overwhelming period for both Cryptocurrency and Blockchain. Both industries covered front pages and made headlines from meme tokens or coins to bull run/upward trends and Non-Fungible Tokens. The last half of the year also generated a major trend known as the Metaverse.

The Metaverse mission of collaborating real-life social interactions, immersive technology, and work has appealed to the public reasoning. Although the metaverse universe is still rising, cryptocurrency is playing a major role already. Let's go into a few of the blockchain network project that is aiding in developing this recent digitalized future.

What is the Importance of Blockchain and Cryptocurrency to the Metaverse?

To fully grasp the significance of cryptocurrency and Blockchain for and to the metaverse, we will briefly go through what the Metaverse means.

The metaverse is an online world that can be explored through three-dimensional avatars and is connected. Metaverse users are allowed to design, work, learn and socialize at a go. You can see it as an internet revolution. PayPal and Debit or credit cards payments are used for the Website, while the Metaverse universe has cryptocurrency to assist in the creation of a digital ecosystem.

Blockchain technology has been an important technology for six major aspects of the Metaverse: transfer of value, digital proof of ownership, interoperability, governance, accessibility, and digital collectability. Blockchain innovation offers an open and moderate cost; thus, making it a perfect match for the metaverse space.

Blockchain Network and Cryptocurrency Projects

Blockchain technology introduced value propositions to the metaverse, which provided sufficient evidence regarding the future increase of cryptocurrency projects and the metaverse blockchain. The technology of blockchain offers a safe, transparent, and affordable answer for being in alignment with the major mission of cryptocurrency projects and metaverse blockchain technology which has great potential to be maximized.

What is The Sandbox (SAND) Project?

The Sandbox project is a blockchain-based game wherein the players or users can explore a virtual space that contains Non-Fungible Tokens (NFTs) user-generated surroundings, including others. The Sandbox was launched in 2011 as a mobile phone game. It transformed into a challenging game in the Ethereum network with Ether (ETN token) and the SAND token to influence Sandbox's in-game economy.

Sandbox players design their avatars, digital means of identification, and a major concept for the metaverse universe. Avatars are usually connected to a cryptocurrency wallet to handle the non-

133

fungible tokens of a player, SAND tokens, including additional blockchain network assets. The players are also allowed to create virtual services, digital goods, and games by using the Game Maker tools and VixEdit application.

These applications are powerful products that develop complicated and expert video games valuables that can be converted into Non-Fungible Tokens. Because Metaverse users can exchange the assets, it became a Play-2-Earn example for users to generate additional income by playing a game in The Sandbox.

What is Decentraland (MANA) Project?

The Decentraland is a three-dimensional universe for players to design their parcel of land, host concerts or events, generate content, and be involved in numerous social gatherings and activities. The primary Decentraland economy is based on Blockchain technology to create digital identification, ownership, and rareness for special goods. Decentraland was among the popular and well-recognized Metaverse projects dominating the huge metaverse buzz in the final quarter of 2021.

This project, Decentraland, is among the early pioneers in the metaverse. It is a three-dimensional universe, and one thing this does is enable players to own parcels of virtual land as they participate in various activities in the space. Decentraland players can also organize events or concerts, participate in social engagement and generate content on the environment. For your information, the Decentraland project was running on the metaverse before the hype began.

Decentraland was launched in 2016 by Ari Meilich and Estaban Ordano. They used a normal two-dimensional game and changed it to a big world that has Non-Fungible Tokens valued at hundreds of dollars in their thousands. The Decentraland project has its ERC-20 token for utilities known as MANA. How then does Decentraland belong to the metaverse?

It fulfills various features of the metaverse, such as an in-game event, a three-dimensional interface, social elements, and a digital economy. As extra projects get connected to Decentraland, it developed a part for metaverse hub. Another reason Decentraland is popular is because of its virtual land asset or real estate non-fungible tokens known as LAND.

In addition to offering the power to vote in Decentraland DAO (decentralized autonomous organization), LAND experienced a tremendous rise in its cost; hence, it is more famous amongst other investors and traders.

What is Enjin (ENJ) Project?

This Blockchain project platform concentrates on the generation of Non-Fungible Tokens used as items for in-games. The Enjin project has launched an SDK; software development kits to enable the generation of Ethereum-based non-fungible tokens easy for the common user. Since non-fungible tokens are significant to the metaverse universe,

Enjin sought a safer means for the tokens to be minted by people. The major challenge many people have with non-fungible tokens is their ability to be illiquid. Thus, looking for a person to buy your non-fungible tokens is difficult and time-consuming. Although, an Enjin non-fungible token can be exchanged for ENJ coins by being melted.

This implies that non-fungible tokens value will never cease in as much as the cost of ENJ tokens does not get to zero. As there is no use waiting for a buyer to purchase the tokens, changing non-fungible tokens to ENJ can offer liquidity instantly. Enjin is likely to be an important part of the metaverse universe in assisting the scarcity and supporting digital collectibles.

What is Bloktopia (BLOK) Project?

Bloktopia is yet another virtual reality metaverse universe game fixed in a skyscraper with 21 floors. Like The Sandbox and Decentraland projects, Bloktopia desires to become a hub for work, events, socialization, and many others. The21 floors skyscraper in which it is fixed stands for 21 million BTC, which is the maximum supply of Bitcoin. The Polygon blockchain supports the Bloktopia project to give support to its four primary sectors:

-Learning; Bloktopia will serve as a means for its users to get knowledge concerning blockchain and how it aids in powering the metaverse. Also, it offers an additional means of accessing and interacting with cryptocurrency.

-Earning; Bloktopia is also embracing the play-to-earn model alongside its native currency, BLOK, a virtual land asset called Reblok, as well as Adblok, an advertising chance.

-Playing; The users of Bloktopia can interact with and communicate with their friends and family online then have access to a large range of content and user-created games.

-Creating; Blocktopia makes the tools needed by gamers available to develop environments, including digital advertising spaces.

The subject of play-2-earn has become a trendy subject in the metaverse gaming sector. The thought of making an income via playing and participating in a game is very appealing in that it creates a natural impact on one's mind. When the Sandbox and Decentraland showcases an easy means of selling land and real estate, on the other side with Bloktopia, it goes further. Each Reblok floor can be given out for rent or hire by tenants or an event. Bloktopia users can generate revenue from ads as others on their level spend time.

What is the Star Atlas Project?

Star Atlas is the last and most likely to be the Metaverse cryptocurrency project with greater innovations. This innovative metaverse game is developed on basic features such as a Decentralized financial system and technology, a Blockchain network, real-time experiences, and graphics including multiple-player video games.

How Star Atlas connects the metaverse universe and the blockchain technology is seen on the blockchain known as Solana, which is the Star Atlas foundation. The metaverse gaming system of Star Atlas enables its users to purchase digital valuables, also called assets such as ships and crew, tools, equipment, land, or real estate.

Furthermore, the Start Atlas project also collaborated with POLIS, a monetary system for in-games whose purpose is to serve various in-game processes majorly. It is easy for Star Atlas to soar swiftly with the top-ranked metaverse cryptocurrency projects in years to come due to the several functions and experiences worth looking forward to.

The development of the metaverse universe still has a long path to reach its peak. A good observation of the various projects will make you realize that the basics, look, feel, and mechanics are simple. Several of them are still planning and preparing and are not ready for a test run yet. One thing is sure, though, the number of projects is still increasing. Whether it is a small gaming industry or a large metaverse cryptocurrency project, what matters is that development is occurring fast.

However, one thing for certain is the amount of developing metaverse-crypto projects is adding up. Regardless of the kind of company investing in the projects, big or small, gaming or real estate, the metaverse space is rapidly developing. The projects just discussed are among the first and most popular to be launched in space. Others will be launched, and you have to be on track.

The Relationship between Cryptocurrency and the Metaverse Space

If you don't know the trendiest technology item, you are missing out on some real juicy treat. Metaverse is the steam giving out hot, sizzling opportunities to save you the trouble. Facebook's recent rebranding and the sporadic increase in value showed by the Metaverse projects and tokens is a piece of key evidence that metaverse is feasible. To add the icing on the cake, the metaverse blockchain network projects provide just the perfect vibes for incorporating Metaverse and blockchain technology as one.

Although Metaverse is obviously in its first stage of development, it is remarkable to see an increase in Metaverse projects progressively. Blockchain technology offers the important features of great use in the metaverse: ownership, security, decentralization, and Interoperability. Is the metaverse-blockchain technology relationship possible to blossom in the nearest future? How exactly will this relationship help the metaverse?

Why are interoperability and commonness critical in the Metaverse?

A significant quality of the hit movie Star Wars that was so striking was portraying a finished future. Other movies that depicted the future showcased a neat and high-profile world, while Star Wars changed the narrative by presenting an unclean and used future. We want something relatable, and this natural or common feel is key even regarding the virtual environment.

Something about the physical realm we live in is that its uniqueness and signatures are embedded in it. This world is a stage that accommodates human interaction, ethics, and culture, making it appealing to our psychology for us to involve it. Besides moving, getting along, interacting, and socializing with various people, including strangers whenever we please, our physical presence in this world comprises of our physical features like retina, voice, fingerprints, as well as global characters like birth certificates, signatures, passports, including vital documents like credit or debit cards, bank accounts, house or land documents, and others.

Also, who we are (Identity) is shaped and widened by what we produce and consume, such as movies, music, and books that we like, are connected to, and express what we feel better. What we understand about the real universe can also be modified. For instance, while asleep, who we are is constant regardless of our active or inactive senses. Likewise, we may face stolen, broken, or damaged identities, yet you always know there is me, the true me. An effective three-dimensional virtual space will incorporate some of these elements.

- Easy accessibility and availability- When in a traditional setting, being at the right location is important for unrestricted access into the virtual realm. A desktop computer, headset, internet connectivity, and serenity are also important. This realm is different and farther from what we can see at the spot. However, this realm is getting modified due to other modernized tools and smart devices.

The Intriguing metaverse world can be accessed even while movement is made, resulting in an easily available place. The audiovisual devices have been upgraded, such as Gyroscopes, accelerometers, front cameras, and multiple screen touches that have boosted the accessibility of virtual environments. The rise of wearable innovations is still increasing, so this technology would still be easy to access. With the rate of advancement and adaptability, in no time, just blinking the eyelids or nodding the head will take you into a virtual environment that is as tangible as the physical environment we live in.

- Reveal unique personality and presence - There is a digital character that represents every person in the digital environment. Our credit or debit cards, bank accounts, and membership identification in different establishments serve as authentications for our representation. These documents can be used in every location to carry out financial, personal, entertainment, or socially related transactions.

The increase in these systems affects data flow modified from the Producers-Consumers then Prosumers; a set of people who develop, observe and change content. So, a persona or personality is about an individual's total online availability, the content he created in digital space, viewed or modified. The virtual environment is also in need of various online presence.

Several virtual environments also have bank accounts with funds or currencies in the virtual account that enhances the user to buy, create and view. These users are usually given different identities to have access to certain content. With these systems in place, watching a live band in a faraway distance or country will be possible.

Interoperability

The element interoperability is the pushing force of the Metaverse. Interoperability is a simple term that means accommodating different systems or operations in one place; inter-operation. When the internet is observed, it has various layers that enhance multiple and various interactions between one another; networks and their subnetworks alike.

In reality, as we travel and visit different countries, our identities that reside in our bodies move easily also without challenges. Something keeps the continuity in reality even while we transit through different locations. So, this ability hopes to be achieved in the virtual world also. With this element in place, the many virtual environments available would coexist and operate in an interchanging manner to be among a bigger picture.

Factors that boost interoperability

The initial factor common for the virtual environment is VRML, known as Virtual Reality Modeling Language, which is described as a single three-dimensional environment as an individual world whose file is easily downloaded and portrayed using a virtual reality modeling language supported browser.

After the virtual reality modeling language is the X3D responsible for expanding the graphical contents of Virtual Reality Modeling Language. Next is COLLADA; COLLAborative Design Activity, and interchanging format. COLLADA's popularity made sure goods and services were traded swiftly between virtual environments. The COLLAborative Design Activity is also generally used for third dimension objects interchanging techniques besides the virtual world.

Layers of Interoperability

There are several layers to which Interoperability exists in the Metaverse space. First, a standard model should be available to multiply geometry, assets, properties, and acts of virtual spaces. These model standards include Virtual Reality Modeling Language, VRML, X3D, and COLLAborative Design Activity, COLLADA.

Defining the Immersive Reality Experience

Fantasy is a world that has always been associated with the female gender perhaps their love for fairy tales and happy endings leave them smitten. However, every human has experienced fantasies at various points in life. Fantasies are dreams, what we aspire to do, spend our time daydreaming about, or wish could happen.

Not everyone will have their dreams come true, but with simulators, that child who'd love to dress up as an adult or an aged individual driving his dream car in the metaverse will be a dream come true. Immersion is when we move from a real environment to an imaginative world. Anything that shifts your attention from the real to the fictional world is immersive.

The Movie Industry and Immersion: several people have attested that a movie is a relief that helps them travel into another world different from reality. A lot is explored and seen in movies from the highlands, waters, forests, and others. Movies assist you in being present in any geographical location of the world or industry, be it Korea, football stadium, Mississippi river, etc. Watching movies cannot provide you mind-blowing experiences albeit, the Immersive feature of Metaverse will do.

Digital Ownership; A New Concept is Birthed

The concept of Ownership is now new. It is aged as humanity, an ancient element. The ideology of ownership is also tried by metaverse space, the other ever active and consistent world. Non-fungible token's economic increase in popularity shows the extent to which people value ownership and desire its replication in the metaverse.

Metaverses are codes in origin. The introduction of the metaverse world portrays a great mental transformation of what digital presence and features mean in as much as they are not new concepts and the video games. Second Life of 17 years shows that recent games portray this: Roblox, Fortnite, and the Sandbox, a game for users to purchase digital land, design, and make money from their assets on the blockchain network.

Yes, we have defined the Metaverse world as many things, yet its nitty-gritty is being a code. It is a code of zeros and ones upon which an enormous amount of data is laid, a designed surrounding filled with synthetic assets, developed and felt from our innermost parts. This world, with its beauty and possibilities, originates from a code. Everything existing in the metaverse can only be after a code has been developed, from the clothing and accessories of avatars to the avatars, real estate, fashion houses, and all others.

The legal perspective shows that immersion in the metaverse universe will challenge several legal ideologies that arose from the physical environment and the basic foundation of ownership is part of this. Significant questions that will affect the Metaverse like digital assets pass for being owned or new types of ownership eventually arise from the virtual world. Questions that will catch the minds of Metaverse users and legal practitioners and makers as virtual worlds become a part of the world.

The Metaverse of Property and proprietary

Properties and Ownership is a concept legal in every form and ancient. Ownership is a concept that prehistorians say its beginning was around Neolithic life of sedentary including agriculture that produced this idea of owning and property. It is an ideology whereby capitalistic worlds are still being operated by today.

Different ownership rights in land assets, shares in a corporation, musical performances are examples that offer monopoly of resources gets to the beneficiaries. The identification of monopoly is understood publicly to generate from offering incentives for the owner to invest in boosting the property since it obtains gains from its sale or usage. Similarly, a proprietor or owner can flex exclusive rights or power over an item.

Copyright, also recognized as Intellectual property, was developed to allow the same reservation of ownership for the beneficiaries involved. This concept is not new to the establishments developing the metaverse. Other architectural and entertainment companies of Metaverse use IP or Intellectual property to secure and make money from the investment.

Also, a transparent incentive, especially for the establishments to develop a virtual environment on proprietorship whereby everything designed is qualified for intellectual property security, graphic characters, elements, and features. This world of metaverse has the words Intellectual property heard at every turn will pose issues and intriguing legal problems for metaverse users whose expectancy may not be realized in the Metaverse. So the people reason that if they have ownership, in reality, how much more the metaverse space.

Ownership versus Licenses

Ever since the invention of the internet, many cases concerning land have described how the digital products of users keep them desiring that they be produced in the exact manner they are conversant within reality.

One of such cases was held in 2012 by a company known as Usedsoft1, where the Court of Justice in Europe had a debate concerning legal rights for software buyers to resell their already used lIcense on the market for second hand got the concentration of the virtual world totally; is it possible for software licenses to be sold again? Or novated instead?

Another case was witnessed in 2018 by Capitol Records against Redigi2, and the Court of Appeal in the United States was questioned similarly for Second Circuit regarding users who would like to give out digital music files they obtained legally to purchase used virtual music files at a half its price on iTunes from others. Not too long ago, Torn Kabinet3, a Dutch company, took their case to the Court of Justice for the European Union to make efforts in obtaining authorization for e-books to be resold legally at second-hand value.

What will these cases result to? These digital products sought to be sold by second parties were not created by them. Digital e-books, software, digital movies, and music files cannot be resold at a market for second-hand goods because the Individuals who bought these goods are not their owners but got licenses for those products to be in their possession.

Tangible goods have two different forms by which they can be authorized:

-Property of Tangible products in the form of papers, docs, plastic boxes, and others.

-Intellectual property is also known as Copyright in the form of film, book, software, and music.

In comparison to tangible properties, Copyright or Intellectual property are authorized only by individuals assigned through the law as benefactors from IP, which is the authors. When a material feature of a wok is lost, like when books or discs are changed to nothing beyond a file, the file does not contain equal digital ownership, which gets obtained individually from the copyright. The contents of a digital file are data in zeros and ones, while information otherwise, data cannot be fixed exactly as physical items. Just as ideas, data, and information are freely flowing.

All the above-described cases show the consistent pressure between the desires of digital products users and the establishments that license them. For the Usedsoft1 case, the Court of Justice for the European Union did not cancel out one choice: the chance for second-hand licenses to be transferred. According to narratives, it would likely be unjust to discontinue the second-hand market existence, and it will be a hindrance to the rights of the consumers, which is why the court tried to reach the ground in between.

Presently, the issue that keeps resurfacing is the undue restraint of consumers. At the same time, the owners of intellectual property rights have successfully managed so far the thought that digital services ought to be traded. As to goes on, convincing metaverse users their assets are in existence by a restricted end-user license will be more difficult. Just like in reality, users will love to claim ownership over a virtual land, car, or handbag they put based in the Metaverse space.

This is simply resolved by making the users understand that the digital items on the metaverse are licensed. It sounds simple and inviting but is unrealistic. Digital products users see limited licenses as a poor replacement for ownership. Using many socioeconomic theories, this concept is illustrated. The concept of human connection to ownership is demonstrated and a concept and the endowment effect. Based on this theory, people would generally fix a greater value on items they own than the worth they'd fix on objects that are not theirs, say if they obtain limited authority over it.

Because of this theory, we can understand why digital products will never be publicized as licensed commodities but rather are shown as sold to users. This means ownership gets sold, while the case is different for licensing. The truth is the virtual world has nothing for sale; it is a huge paradox every Metaverse user and developer has to face squarely.

Entering Non-Fungible Tokens

The progressive increase in non-fungible tokens' popularity shows the extent to which humans crave ownership and the extent of the appetite to which they want it replicated in the virtual world.

The idea of Non-Fungible Tokens is simpler but ingenious at the same time from a legal perspective. So, if a digital object cannot be owned since it is free data, then another thing that can be owned individually from the intellectual property would be sought. For instance, a certificate of authentication that cannot be falsified is connected to a digital product.

The authentication of certificates given by the product's creator in limited amounts is a wise way for scarcity to be reestablished because a feeling of ownership is gotten and value also so there's no need for intellectual property rights to be transferred or assigned to the person receiving the token. Metaverse users need to understand that a connection to the digital product is what is being sold, traded, or exchanged, and is the sense of ownership sought after regardless of whether it is just a certificate.

Non-Fungible Tokens sparkles all the way more where the AG old concept of ownership is displaced, reinvented, renewed to an intangible certificate from a tangible medium such as tapes, books, discs, etc. Do you know that UsedSoft suggested notarial authentication certificates to let the resale of software licenses in 2007 be logical and notable, a sign of future occurrences?

Currently, the second-hand marketplace for digital products rejoices with others as the metaverse industry and others connected to it is invigorated by its concept. The two major foundations of auctioning, Christie and Sotheby's, are selling non-fungible token works with so much enthusiasm they have not been in the auction houses at all because they couldn't be made special or felt. Since Beeple's, to the First 5000 days, drawings made by Andy Warhol digitally prohibited from auctioning are now tokenized and entering the art market remarkably.

The Reinvention of Digital Ownership

If the Non-Fungible Token seems to solve many challenges that arose from the chase to allow unlikely property rights over digital products, in addition to creating a form of crafty resale technique to permit the main owner of non-fungible tokens to partake of the created by every resale, we would have a greater question to answer that is if our expectations of ownership will be accomplished with it.

Non-Fungible Tokens do not transfer or authorize a monopoly I have a digital product or grants permission to its holders to choose the usage, distribution method, or presentation of the work. When we displace the previous excitement that comes from being a monopoly art collector, removed from the object and transferred to a certificate instead, our comprehension of the concept of property and ownership is being affected. This paradigm shift relates a thrilling feeling about our values as humans. It is time for meta-propriety or ownership.

Building the Metaverse: Technology is experiencing recent advancement. The launch of upgraded online platforms has enlarged the possibilities for public and private space in the virtual environment to be thought of and developed.

Digital Collectibles: the building blocks of the metaverse

Digital collectibles are items built and allocated on a blockchain network to offer collectible ownership and transparency and are issued in limited amounts. Non-Fungible Tokens are rare digital collectibles. The technique of digital collectibles allows creators to attach value and scarcity to these objects so users can possess assets and apply them in creating digital environments and experiences that have real-life value. The launch and increase of non-fungible tokens spurred the development of industries. They raised the demand for top-quality digital assets for users in existing and virtual digital environments.

Using Digital Collectibles

Many tech companies and even other industries are integrating with leading three-dimensional artists, fashion designers, and artisans in the real world into virtual reality to create metaverse collectibles for virtual users and own as many quantities as they can afford.

The Current use application of digital collectibles within the metaverse space include:

-Video games: digital collectibles are supported by several blockchain games like Enjincraft. With digital collectibles, you can experience a thrilling digital adventure and then create your stories based on these experiences.

-Virtual Reality social platforms: Many digital collectibles are also compatible with the open world and Sansar virtual reality platforms. Sansar is a virtual environment that allows users to develop individual space and playful worlds enriched with non-fungible tokens collectibles, including other virtual assets.

Governance

Many questions concerning the administration of the metaverse and if Facebook is in charge of its operations are being asked since the change of its parent name to Met and huge financial investment into the Metaverse. In 2010, an economist commented on Facebook, the social media platform beginning to look like a nation-state when the relationship and external operations were considered. At that time, the social platform had not reached a user base of 1 billion, but the meeting was held amongst them; Mark Zuckerberg, Chief Executive Officer of Facebook, and the Prime Minister of the United Kingdom, David Cameron, seemed to be diplomatic interaction.

As the years passed, analysts brought suggestions repeatedly that the population, size, and impact of the social application, Facebook implied that its guidelines could be defined by itself and thus, governed just like an independent nation. An example is when Facebook banned the Myanmar military; this showed a sign of sovereignty and superiority. Presently, Facebook has rebranded itself to Meta and is set to build the Metaverse; the exact issue could occur and be applied to its virtual reality offerings.

What the metaverse is seen as is a virtual world in unity that mirrors the real world. However, the real and virtual worlds already have boundaries that make them very complicated to rule. If this is the same for social media regulation and governing, what makes the metaverse any different? So, the statement, who will rule the metaverse is an important question that should be answered in the early beginnings of metaverse implementation.

What Is the Metaverse?

This metaverse world is a three-dimensional virtual environment where people can communicate with their environment and others.

The metaverse has three properties that define it wholly:

a. Interoperability

b. Universality

c. Social environment or surrounding

Also, the metaverse is founded on their primary technologies:

a. Extend Reality or XR

b. Blockchain technology

c. Artificial Intelligence

As far back as the 90s, the theoretical concept of metaverse has been in existence; however, we did

159

not see any practical advancement towards that till five years ago. When blockchain technology was advanced, virtual reality became accessible, and artificial intelligence ready-to-use packages became available, implying that industries and companies could begin developing their versions of the metaverse.

When Interoperability is absent, it removes the authenticity of a Metaverse because they become independent apps that have a single or more virtual environments contained in them. An example of such is Decentraland and The Sandbox, which can be referred to as a metaverse for games only.

The vision of Facebook is different from the others as Chief Executive Officer Mark Zuckerberg stated that the concept of Meta company metaverse is decentralized, uncontrolled by any establishment referred to as Embodied Internet, at 2021 Keynote. The metaverse-cryptocurrency projects The Sandbox and Decentraland are decentralized financial systems also via Blockchain technology; however, they are smaller and have additional limited scale. Facebook already has a thriving presence on social media of about three billion users, unlike the other projects to design a community-based and universal metaverse system.

What Does it Mean to Govern the Metaverse?

There are no set down rules and regulations when it comes to governance, especially in the metaverse as a sole proprietor or company will not regulate it. Both developers and creators will develop the metaverse to create digital products and new experiences that unblock a large economy and is interoperable compared to the virtual spaces tightened by the policies and different platforms of nowadays. Zuckerberg also stated that Meta, formerly known as Facebook, will be an enabler and booster.

Facebook could also be said as the same since it depends on content generated by users and publishers in other industries to push and generate traffic. Even on Facebook's platform, the developers and creators are both responsible for shaping users' experiences yet; they go through the algorithms and internal operations of Facebook. Because of Mark Zuckerberg's unclarity regarding administration and governing, his company recently fixed up an Oversight Board panel to cross-examine its contents' policies.

Any potential and pending challenge connected to the governance of metaverse has to be handled on time to ensure that the same pattern does not occur on a different or newly launched platform. It is a good step and relief that several public bodies and governments are getting into the metaverse to set up administration.

Governmental Bodies in the Metaverse

Facebook has huge plans from the metaverse; it is no longer news or story. The world is gradually getting in touch with the concept of the Metaverse and its possibilities. Governmental bodies have taken note and playing their cards seriously as well. If the metaverse will cause an impact on the world, economy, and way of life, then governing authorities should partake. Some countries have invested in partnerships at the early stage. With this action in place, envisioning a great administrative structure that caters to the future is in place. The following countries have set the pace:

South Korea and Seoul

South Korea is amongst the nations with the highest population of gamers; it is the fourth-biggest world gaming market. There are several metaverse players in places like Zepeto, island, Netmarbke, etc. The South Korean government is really impressive as they do not only support the metaverse world and technology. ensure that the South Korean Ministry of Science including its ICT center raises metaverse alliance to regulate the efforts of Metaverse. Over 500 companies have joined the metaverse and in Seoul, the city, plans have been made to design a municipal metaverse space.

Tencent in China

A quarterly call for earnings was made recently, and Tencent made a public announcement that the metaverse is welcome in their region but will make use of the laws and regulations of China. In as much as the Chinese administration will back up the metaverse development, user guideline and experience must be in obedience to the Chinese government rules. China is a nation with a rooted track record of technology marketplaces crackdowns; therefore, the metaverse will be centralized there (controlled by the government).

The Barbados Metaverse Embassy

Barbados is the first sovereign nation worldwide to develop an embassy in a metaverse space. It formally recognized virtual reality spaces as governable or administrative entities. The Barbadian Government is also collaborating with several metaverse companies such as Somnium space, Decentraland, SuperWorld, and maybe Meta, so an undeniable presence will be created in the governing landscape platform.

Is the Metaverse Future an Open, Non-profit Regulatory Body?

The above countries and how they intend to partake in metaverse shows that the governing of metaverse needs to be taken seriously by the world. However, China has no straightforward administrative structure on the ground. Normally, a worldwide alliance for metaverse should be gotten not with 500 but many thousands of members to ensure the metaverse functions without third parties no use for regulatory bottlenecked bodies or government.

A decentralized structure of the metaverse universe implies that daily activities will be responsible for running themselves without interventions from any party. This transparent, non-profit collaboration will intervene only through lawmaking and high dispute-solving occurrences.

Are Cryptocurrency Laws Going to Serve as the Basis for Metaverse Governance?

The current task at our doorstep is to subject blockchain technology and cryptocurrency to the regulations and to govern to ensure that everyone is ready for the future with metaverse. The approach to cryptocurrency still very much differs between Governments and territories and may pose an issue since cryptocurrency is the basic foundational block for the metaverse and its economy. Everyone must reach a consensus concerning the three key technologies backing the metaverse before building on it to create a path for future regulations.

Web3, Interoperability and the Metaverse

Web3 is the Metaverse for several people, while others see it as a mysterious world, intertwined with so much cryptocurrency that newbies sound strange. The metaverse is a real-time internet based on activities. Real-time activities have been in existence for a while, some of which include games and video conferencing. With metaverse, the digital perception will be highly elevated, and new types of

165

creativity and individual expressions will be formed in readiness for new memories to be created.

Does the metaverse need Web3?

Why Web3?

First, we have Web1, the initial World Wide Web founded on Open-source like Linux, Open standards like HTTP/HTML, and permissionless development like Personal Computers software. Many of the existing biggest internet companies such as Google, Netflix, and Amazon were created on the Web1 ecosystem while others gained from their expansion into it like Apple Incorporation and Microsoft Corporation.

Then, Web2 is greatly focused on content created by users such as social networks, blogs, wikis, etc., and asynchronous. The majority of the Web2 ecosystem was also created on the technology of Web1 or gained from search to acquire audiences. The biggest Web2 establishments were created on similar transparent and standards-based surroundings that boosted Web1 yet built a wall garden ecological system to allow social connectivity and content development.

Significant examples in this category include Facebook, now known as Meta, FAANG company's youngest establishment, and Google-owned YouTube in creating wall gardens for networking and content creation by users. Roblox is a recent company in this category.

The significance of walled gardens is making things easier to achieve and providing access to highly populated audiences; hence, it's a success. However, these gardens are environments that need permission and control the activities in its surrounding, including an expensive rent rate for Access. Although many companies built within the ecosystem of Web2 were successful, such as Twitter and Facebook, no company has emerged with a stable, very scalable business with over a trillion dollars market capitalization within the Web2 gardens.

We will go through the characteristics of the Web3 ecosystem for changing the paradigm.

- Value Trade or Exchange

- Self-Sovereignty

- The Internet Re-Decentralization

- Value Trade or Exchange

The exchange of value is offered beyond data exchange. It is a transforming feature and if you forget anything else, keep in mind that Web3 is transformative and boosts the exchange of value between apps. The technology that enables value exchange on blockchain networks is smart contracts.

What is the Blockchain Network?

This is a distributed ledger that permits apps, governments, companies, and communities to exchange value such as currencies, property, assets, etc. in a transparent and programming manner with one another without the use of intermediaries, brokers, or custodians, which cause a constraint on innovation as they extract enormous rents.

On the other hand, blockchains on smart contracts allow great creativity instead of the innovation model, hub-and-spoke that allows applicants to move amidst permitted ecosystems. The skill of exchanging values in a programmed manner between various parties is transforming even for the civilization of Manas the original and initial internet, Web1 was designed. This is just like moving from an online medium regulated and handled by an individual to another platform with every creativity embedded.

In a podcast, Balaji Srinivasan stated that we often perceive communities, companies, countries, institutions, etc., as various things, but all these will become social network projections in the future. The social network in this context is not Facebook and the likes, but the networking connections brought in from various contexts.

The enabling software, Blockchain, can only achieve this. It offers Interoperability for storage, exchange, and setting up property rights or ownership, assets, identities, and currencies. In this ecosystem, the companies and communities turn into software apps.

- Self-Sovereignty

Without doubts, you logged into Facebook or Google login to communicate and socialize with a huge amount of online apps. These login apps are databases for users operated by big, centralized bodies. Changing or transforming the model is a critical aspect of Web3. In the model, the identity of users will be owned by them and decide the apps to be in touch or interact with rather than have your identity owned by a company and giving you permission for other apps.

This is achieved through digital wallets like MetaMask for blockchains compatible with

Ethereum and Ethereum itself or through Phantom, which was applied on Solana blockchain. Your means of Identification is your wallet. It can give you access to multiple apps on the website that require interaction with both properties and currencies you own. A large number of a dependency across the websiteDeFi also, Decentralized Financial apps and metaverse memories facilitated from game items, interoperable avatars, self-expressive objects, etc., is received by the project's owners and creators.

What is Web3?

From serverless software to distributed blockchain processing

The World Wide Web was a serverless software that is partly a misnomer. This software has a development strategy that permits storage, including additional server-command functions in the cloud, thus taking the burden off builders to think of DebOps, deployment, backend scalability, and several IT features connected to full-stack web software building or development. A lot of Web3 software projects are developed on serverless backends. Using this scenario, Moralis offers an SDK that allows app developers to create web-based software that collaborates digital wallets with server-command, serverless data storage systems.

Serverless forms of architecture operate on SaaS frameworks functioning inside centralized architectural bodies. Moreso, serverless architectures carry the potential to cause the transition into a total decentralized future because:

-Serverless architecture removes the role of operations, availability,designing-in, and scaling from the software functions such as the microservices, which causes functions, e.g., gaming rules, payments, world state, social features, etc. In addition, a distributed computing surrounding can be opened to support the microservices that give extra functions to the newly created ecosystem or, instead, add computing nodes that offer server-command functions.

The advantage of doing this is the reduced dependency customers or users will have on centralized sellers/vendors, including the price optimization of offering these services based on what effective auctions for most available or least expensive computing will be available in comparison to selecting one cloud service host like AWS or Azure whose bidding on computing horsepower will be welcome on the ecosystem also.

The distributed computer nodes will make use of on-chain means for solving guidelines and data storage.

Serverless architectural companies that offer serverless backends will be offered chances in the future but, will they be decentralized?

It is most likely a shift in the business model for industries and establishments that will cause a great hindrance for most of them although, those who create a way to reach there, new start-ups built to be on Web3 will have chances to make great value added to the whole ecosystem.

The Web Assembly and Web3

A recent standard built to permit byte code to get delivered straight into your web browser is WebAssembly. You can see it as a method of directly transporting applications to your browser instead of relying on application store ecosystems. Instead, these applications will function wherever a web browser functions like virtual reality and augmented reality headsets, computers, and smartphones.

WebAssembly is not required to be restricted to web browsers because it can function on backend servers for new classes of flexible server codes.

WebAssembly (WASM) and Containers are properly allocated to benefit from the new shift of decentralized and distributed computing due to high portability, maximum flexibility for a workload type that can be operated and bundling their dependencies. WebAssembly created on Web3 structure of a programming value-exchange, Self-Sovereignty identity and decentralization possess what it takes to radically modify the landscape of multiple apps storage ecosystems.

Disruptiveness of Web3

Web3 takes over the requirement to build or incorporate many services like some related to identity or asset custody which is costly time-taking. It leads to reliance on centralized platforms that may not align your interests with theirs. Hence, many app developers can design their software with small teams, receiving incumbents based on those who developed their companies around costly techs or relied on software partners that need to go ahead and negotiate.

Bundling, Unbundling, and Rebundling

The Chief Executive Officer of Netscape, James Barksdale, said one time the only way he knew how to get money was through unbundling or bundling.

173

Unbundling, for instance, is how iTunes unbundled single songs from their albums while Spotify came along to make songs rebundled into playlists. Presently, most video games are greatly bundled; however, several video games will get unbundled and bundled in the time of Metaverse internet.

Web3, Avatars and the Metaverse

Digital identity is our identity (who we are) and how this identity is expressed online. This means of identification will be more significant than it is now. Our avatars move from one experience to another in the metaverse, mainly depending on an interoperable frame. An example of unbundling is an avatar's properties; recent forms of experiences online such as music, theatre, games, and other apps will be a form of bundling this means of self-expression.

For a great range of apps to be enabled, an interoperable identity frame and property will be needed, which is amongst the uses of Web3 in the metaverse space.

In addition, Web3 offers a framework for creativity for the upcoming generation of P2E (Play-to-earn). Currently, Play-to-Earn is translated to some kinds of cryptocurrency money farming or power-finding like Axie Infinity. Many Individuals had played to earn money since before the cryptocurrency or Blockchain added it up through live streams on Twitch, eSports, or other modding games formats. It is a known fact that players in their large population of millions seek to transform their hobbies into a source of making money, and this is all part of the bundling and unbundling that is occurring in gaming systems. Play-to-Earn will be used to cover up creativity, performance, live-roleplaying, competition, level-designing, guiding, dungeon-mastering, and others. The description between a player and designer will get thinner, so new bundling and unbundling formats will be seen. An interoperability framework for most disruptive economic tradings will be highly needed, and Web3 is just the framework needed.

In a transactional contract amidst a virtual environment, the user and server are enabled by a protocol standard. This offers a test and trial occurrence between the users and servers alike within compatible systems. For instance, Open Cobalt is amongst the productive platforms that allow the creation of collaborative and hyperlinked multiple users virtual working systems, game dependent environments, and virtual exhibit spaces easier since they operate on primary software OS (operating systems).

In virtual worlds, a locator standard aids in searching out locations and major landmarks. This technology is not new to the internet as URLs are one and can also be used in virtual geography. Linden Laboratory has used URLs for their Second Life environment.

Identity Standard: provides special credentials to users that are useful across the virtual environment's boundaries. It can be equal to our real-world passports, social security numbers, phone numbers, license numbers, etc.

Currency Standard: this defines the worth of virtual items and creations, permitting their exchanges—various forms of virtual reality feature in their special virtual currency. In no time, the Open Metaverse Currency will see developments that serve as the globally accepted virtual format of currency.

CONCLUSION

Non-fungible tokens, metaverse-cryptocurrency projects, and the others discussed from Chapters One to Four are well explained to make even novices understand everything the Metaverse entails.

The extent to which Governmental bodies, public and private sectors, individuals, developers, and investors are putting efforts into the operation of metaverse brings nothing but a glimpse of light at the end of this tunnel of efforts. Individuals seeking to go into the metaverse should be aware of its entirety, because just like cryptocurrency, the metaverse may be volatile.

If the metaverse is a fantasy and unreal, what invests put into it real? This is one-of-a-kind technology the world is yet to comprehend and recover from in the long run. Every great technology has its ups and downs. Therefore, it is dependent on the parties involved to thrive in making it successful with time.

Several projects have been put in place to spearhead the launch of metaverse into the world, and they have shown a tremendous increase. Facebook, the social platform for networking, is amongst the main sponsors of Metaverse. Just as was discussed in the chapters above, as proof of its commitment and zeal to the metaverse vision, it changed its parent company name to Meta. A lot is at stake with this recent technology. Its contribution to the health care sector, education, manufacturing, the fashion industry has made lives and the economy easier and more productive.

Blockchain technology is a decentralized system that would aid the operation of the metaverse. Centralizing the metaverse seems necessary because where there is no law and order, chaos is bound to happen, and abuse is inevitable. The scopes of metaverse advocate limitless experiences and originality.

The Metaverse takes us on a tour of the unknown to ascertain our movements therein. Web3 is also important in fulfilling the extension and Interoperability of metaverse. A truly flexible space should be interoperable; just a few individuals are skilled in multitasking. Interoperability helps the metaverse universe to combine various applications and technological upgrades without complications.

For the metaverse vision, this is important to accommodate the large population of people from different continents into one universal space.

As the Metaverse unfolds, we cannot help but wonder how all these will be achievable; however, the uptrends and success of metaverse-cryptocurrency projects remind each one that it is feasible and works with major bands, developers, and companies to put everything on deck to reach its peak in no later time. With this book in your library, you are on the right path to entering the metaverse, partaking of it, and maximizing its potential alongside. Now, we will see where this roadmap leads as we follow from a distance, near or far.